Sensors for Measurement and Control

wn below.

Sensors for
Measurement and Control

Peter Elgar

An imprint of **Pearson Education**

Harlow, England · London · New York · Reading, Massachusetts · San Francisco · Toronto · Don Mills, Ontario · Sydney
Tokyo · Singapore · Hong Kong · Seoul · Taipei · Cape Town · Madrid · Mexico City · Amsterdam · Munich · Paris · Milan

Pearson Education Limited
Edinburgh Gate
Harlow
Essex CM20 2JE
England

and Associated Companies throughout the world

Visit us on the World Wide Web at:
http://www.pearsoned.co.uk

© TecQuipment Limited 1998

First published 1998

British Library Cataloguing in Publication Data
A catalogue entry for this title is available from the British
Library

ISBN 0-582-35700-4

10 9 8 7 6
06 05 04

Set by Peter Elgar, Andrew Parrott and Brett Gilbert
Printed by 4Edge Ltd, Hockley, Essex.

Contents

Preface vii
Acknowledgements viii

1 Introduction 1

2 Basic Principles, Terminology and Characteristics 3
Sensor systems 3
Sensor terminology 6
Sensor characteristics 7
Summary 9

3 Measurement of Motion 11
Linear displacement 12
Angular displacement 19
Proximity 22
Acceleration 24
Summary 25

4 Level, Height, Weight and Volume Measurement 27
Level measurement 28
Measurement of weight or force 31
Summary 34

5 Measurement of Pressure 35
Liquid manometers 35
Elastic pressure sensors 38
Barometers 42
Summary 43

6 Temperature Measurement 45
Liquid expansion thermometers 45
Metal expansion and the bimetallic strip 46
Electrical resistance 48
Thermoelectricity 50
Heat radiation 52
Summary 53

7 Flow Measurement 55
Volumetric flow rate 55
Mass flow rate 58
Velocity measurement 59
Constriction effect devices 61

Non-disruptive flow measuring devices 63
Summary 64

8 Display and Recording 67
Analogue displays 67
Digital displays 70
Recorders 73
Computers and data acquisition systems 75
Summary 76

9 Signal Conditioning and Interfacing: Passive Circuit Techniques 77
Signal conditioning 77
Passive interfacing 81
Summary 84

10 Signal Conditioning and Interfacing: Active Circuit Applications 85
Operational amplifier circuits 85
Summary 94

11 Measurement Application Case Studies 95
Checking the dimensions of a component during manufacture 95
Vibration monitoring on a road bridge 97
Wheel speed measurement in an anti-lock braking system 100
Summary 102

12 Control Application Case Studies 103
Inspection on a production line 103
Control of thickness in sheet steel manufacture 105
Container filling in a chemical process 108
Summary 111

13 Practical Experiments 113
Sensor specifications and familiarisation 114
Experiment 1: the strain gauge 120
Experiment 2: the linear and rotary potentiometer 125
Experiment 3: the linear variable differential transformer (LVDT) 129
Experiment 4: the variable area capacitor 132
Experiment 5: the reed switch 135
Experiment 6: the reflective optical beam sensor 137
Experiment 7: the optical tachometer 138
Experiment 8: the d.c. tachometric generator 141
Experiment 9: the variable reluctance probe 143
Experiment 10: the four bit optical encoder 147
Experiment 11: data transmission 149
Experiment 12: introduction to control 154
Summary 155

Index 157

Preface

Sensors for Measurement and Control is a comprehensive reference book on sensors and their applications. It is aimed at students of instrumentation and control, from GNVQ level 3 to first year undergraduate. It will also be of use to practising technicians and engineers.

The book describes sensor terminology, and the principles and applications of over 50 sensors, including details of display and recording methods and signal conditioning techniques. To aid understanding, it includes several sample calculations, typical sensor specifications, and many drawings and illustrations. Two chapters of case-studies are included to further assist comprehension of sensor systems and practical sensor applications, and to illustrate the importance of sensors in everyday engineering situations.

Although a valuable resource on its own, *Sensors for Measurement and Control* forms part of a self-contained teaching and learning package. The package, available from TecQuipment Ltd, is called the Sensor and Instrumentation System (SIS). The SIS provides a complete course in sensors and instrumentation, and consists of a hardware module with a selection of sensors, and a resource pack which includes an interactive CD-ROM, student and lecturer guides, and automatic data acquisition software.

Each chapter of *Sensors for Measurement and Control* has introductory paragraphs and a summary section. To encourage the reader to think about the subject, and to test the level of understanding, questions for further discussion are included at the end of the descriptive chapters. The book concludes with a structured series of laboratory assignments, which can be achieved on available industrial devices but can readily and conveniently be performed on the hardware of the TecQuipment SIS.

Peter Elgar
TecQuipment Limited
Nottingham, 1998

Acknowledgements

Thanks to John Loasby of Derby University, and also to Gillian Whitehouse and Mike Tooley for their guidance and advice. The production of this book would not have been possible without the help and support of all the team at TecQuipment, including John Lewis, Andy Parrott, Clive Cooper, Phil Sands, Myles Ponsonby, Peter Cooper, Clive Hannaford, Kevin Price, Graham Lewis and Simon Woods.

TecQuipment Limited
Nottingham, 1998

1 Introduction

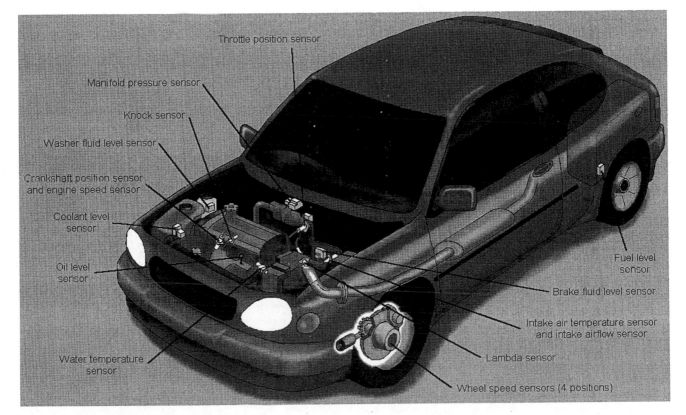

Figure 1.1   Some of the sensors in a car

Sensors for Measurement and Control is an introduction to the increasingly important topic of sensor technology. It is also an extensive reference book on sensors and their applications. This book concentrates on the principles and applications of sensors, including their design, construction, principles of operation, and practical examples of where they may be used.

What are sensors?

Before embarking on a study of sensors we have to be sure what they are. Sensors may be individual devices or complex assemblies, but whatever their form they all have the same basic function. This is to detect a signal or stimulus and from this produce a measurable output.

In this book we will look at different sensors designed to measure various physical parameters. Among the physical parameters most commonly encountered which require measurement are displacement, speed, acceleration, fluid flow, liquid level, force, pressure, proximity and temperature. This is by no means a definitive listing. For example, chemical, sound, or nuclear radiation sensors are not covered. However, the majority of sensors in common use today are described.

The need to measure any of these parameters is usually governed by the industry or utility using them. The precise choice of sensor depends on the nature of the parameters to be measured, and other factors such as cost, reliability and quality of the information needed. Other factors may include suitability of the sensor design to the environment in which they are to be used, or whether the information is needed immediately or for later analysis, at the point of measurement or some distance away. For instance, a temperature sensor used in the home would look different to one used in a chemical plant. The sensor in the chemical plant may be inaccessible, or subjected to high temperatures and pressures or be in a corrosive atmosphere.

Why do we need sensors?

Sensors detect various physical parameters, but it is what we do with this information which makes sensors useful. In general there are two distinct areas where sensor technology is used. These are for gathering information, and for controlling systems.

Sensors used to gather information provide data for display purposes to give an understanding of the current status of system parameters, for example the car speed sensor and speedometer. Alternatively they may be used for recording to provide a permanent account of performance or parameter variations, such as the tachograph used in lorries and trucks, which records speed against time.

Sensors used to control systems are usually no different than those used for information gathering. It is what is done with the information from the sensor which is different. In a control system, the signal from the sensor is input to a controller. The controller then provides an output to govern the measured parameter. For example, the information from the wheel speed sensor in an anti-lock braking system is used to control applied brake pressure and thus stop the wheel skidding during braking.

Both information gathering and control are dealt with in this book.

Why do we need to study sensors?

Technology has advanced rapidly over the last few years, well beyond the expectations of most engineers and scientists. The sophisticated equipment we find in the home, workplace and many other environments includes technologies that less than ten years ago might only have been a laboratory novelty.

The major reason for the availability of such equipment has been the development of computers and microprocessors which are used as sophisticated, flexible and yet low cost controllers. However, the operation of such systems would be very poor, if possible at all, if the computer's decision-making programs were not being supplied with suitable, up to date, quality information about the state of the external system. Once this information is gathered by sensors, it is conditioned to be in the desired format, and then input to the computer system to process and generate an appropriate response. All elements of a sensor system have to be of the required level of performance to suit the quality of control needed by the application. If one element is below standard then the whole process can be degraded.

Figure 1.1 shows where some of the sensors are located in a car. In just this one application there are sensors which measure temperature, liquid level, angular velocity, pressure, and proximity. So you can appreciate the importance and widespread use of sensors, take a few moments to make a list of equipment and applications you use or come across regularly, which you think includes sensors. Identify the physical quantities they measure. Once you have read this book, return to the list and see if you can add to it.

Sensors have become so commonplace in modern society that most people take them for granted. This makes it all the more important for technicians and engineers to have a working knowledge of sensors, to enable them to choose the appropriate device from a catalogue specification, or to repair and calibrate sensors installed in some existing piece of equipment.

Layout of the book

The main body of this book is separated into chapters which detail sensors for the different types of commonly measured physical quantities. The book incorporates drawings and diagrams throughout to illustrate and aid comprehension of the topics

covered. Where useful to the understanding of a particular concept, and where a topic lends itself to explanation in this way, some of the mathematics behind it has been given. This may be in the form of equations, sample calculations or graphical analysis. Where mathematics has been used it has deliberately been kept simple, and has not been applied to all topics because to cover all the theory involved in sensor technology in this way would be impractical.

In practice, the raw signals generated by sensors are often not in a form that can be used and so signal conditioning and processing is required. The techniques for this are covered after the sensors have been dealt with. Likewise, instruments used to display and record sensor outputs are covered to complete the picture.

Throughout the book you will find a number of 'key facts'. These are intended to highlight points of interest, in some cases fundamental to the understanding of a particular concept. They may also be useful for quick reference or revision purposes.

Chapter Two deals with the definitions and terminology used in the field of sensors that are used throughout this book. It defines what is meant by measurement and control systems, the functional elements, typical examples and sensor characteristics. It is strongly recommended that this chapter is read and understood before the other chapters, to avoid any misunderstanding or problems in later chapters.

Chapters Three to Seven concentrate on explaining the operating principles of different types of sensors commonly used for measuring a variety of physical parameters. A chapter is dedicated to the sensors used for motion; level, height, force (weight), mass, and volume, fluid flow, temperature, and pressure. Both electrical and mechanical sensor types are covered to provide a single comprehensive reference source. The advantages and disadvantages of each type of sensor are discussed as well as typical applications.

Chapter Eight discusses recording and display. It explains the difference between recording and display, analogue and digital signals, and covers common methods used.

Chapter Nine covers signal conditioning using passive circuit techniques. It introduces the need to condition signals into the required format. It discusses many techniques and includes explanations, relevant mathematics and example calculations.

Chapter Ten follows on from Chapter Nine to discuss signal conditioning using active circuit techniques.

Chapter Eleven is a study of measurement applications. From the information covered previously in the book, this chapter details examples of where sensors may be used in practical engineering systems.

Chapter Twelve uses case studies of where sensors are used to illustrate control applications. It covers engineering systems from different industries and environments.

Chapter Thirteen is a structured series of laboratory assignments which puts the previous theory into practice. It details twelve progressive experiments, which can be performed on available industrial devices but can be more readily and conveniently be performed using the TecQuipment Ltd Sensors and Instrumentation System (SIS).

Throughout this book cross references are given to link areas of common or overlapping interest. Each chapter includes an introductory section, a summary and a number of questions to test the level of understanding.

2 Basic Principles, Terminology and Characteristics

In this chapter we introduce the concept of sensors for measurement and control systems, and define some of the terminology used.

Sensors and sensor systems may be mechanical, electrical, or a combination of both. They are used extensively in many fields, industrial, civil, military and domestic. They perform tasks as diverse as checking the dimensions of an item on a production line, controlling an electricity generating power station, checking the level of water in a domestic washing machine, and displaying the speed of a car. Because the nature and applications of sensors is so varied, it is important to clarify what a sensor is, and establish the precise meaning of the terminology.

As we saw in Chapter 1, a sensor is a device which detects a signal or stimulus and produces a measurable output from it. Figure 2.1 shows some examples of sensors. The thermistor and the strain gauge both produce an output which is a change of electrical resistance. Many sensors produce electrical outputs, not only relating to the quantity being measured in terms of resistance, but also in terms of voltage, current, or frequency. A spring balance produces a change in displacement, and a pointer moves a distance along a scale proportional to the load on the spring. The Venturi tube measures a difference in pressure to determine the flow rate of a fluid.

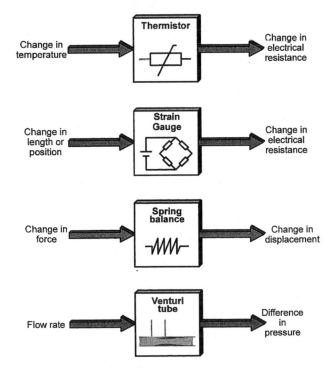

Figure 2.1 Examples of sensors

There are a wide variety of forms of output signals from sensors. The outputs must be in a form appropriate to the display requirements of the system.

The terms sensor and transducer have similar but subtly different meanings, and are sometimes confused. A transducer is any device that transforms one form of energy to another. Hence a sensor is (usually) a transducer, but not all transducers are strictly sensors. For example, a domestic light-bulb is a transducer that converts some of the electrical energy supplied to it into light (and heat). The purpose of domestic light-bulbs is normally to illuminate a room, not to indicate the presence of electricity. However, if we used the same light-bulb to test an electrical circuit, that is, to illuminate when electricity flows through it, it could then be considered a sensor.

Sensor systems

There are many types and definitions of systems, but for our purposes we shall consider a basic sensor system as something which produces a quantified output from a different form of input, by means of some kind of process. Figure 2.2 shows a basic system in the form of a flow diagram. As you will see, flow diagrams are a useful technique of explaining what happens in measurement systems.

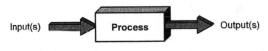

Figure 2.2 A basic system

We can usually classify the application of sensors into three types of system. These are measurement systems, open-loop control systems, and closed-loop control systems.

Measurement systems

A measurement system displays or records a quantified output corresponding to the variable being measured, which is the input quantity. Measurement systems do not respond in any way to the input quantity apart from displaying it to the user in an understandable form. Figure 2.3 shows the basic flow diagram of a measurement system.

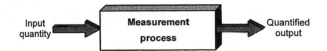

Figure 2.3 Measurement system

Figure 2.4 Functional elements of a measurement system

As an example of a simple measurement system, consider a liquid-in-glass thermometer (which we shall discuss in detail in Chapter 6), used outdoors to display air temperature to say, a gardener. The input to this temperature measurement system is the heat transferred to the thermometer from the air, and the quantified output is the temperature reading in say, degrees celsius, on the thermometer scale as read by the gardener. The measurement process is the conversion of heat from the air to the temperature reading on the scale. This is purely a measurement system because the thermometer in no way controls the outside air temperature.

The measurement process can usually be divided into specific stages, as shown in Figure 2.4. The thermometer example is a very simple measurement process. Here, the sensing, signal conditioning and display functions are all built in to the thermometer and by its nature are an integral part of the device. The signal conditioning is the conversion of heat energy to the movement of the liquid. The display is the scale on the side of the thermometer glass. Many measurement systems are more complex and it is useful to reduce them to components of sensor, signal conditioning unit, and display or recording unit.

In Figure 2.4, the sensor converts the physical quantity being measured into a signal which can, after modification, be displayed to the user in an understandable form. The signal conditioning unit modifies the signal from the sensor to a form which can be used by the display or recording unit. If it is an electrical voltage, for example, it may need amplifying, or if it is a small mechanical movement, it may need transforming into a larger or different type of movement. It may even be a series of light pulses which needs transforming into a series of electrical pulses.

After modification, the signal is displayed to the user, or saved for later use on a recording device. There are many types of display and recording device which show the quantity being sensed, and several methods of displaying this quantity. For example, it may be shown in terms of a numerical output, movement of a pointer over a scale, or printed on a chart or graph. Chapter 8 discusses display and recording devices, and Chapter 9 and Chapter 10 discuss techniques used for signal conditioning.

Open-loop control systems

Both open- and closed-loop control systems try to maintain a variable at a predetermined value. Control systems incorporate measurement systems, but unlike a pure measurement system the output from a control system regulates a parameter whose value is not necessarily displayed to the user.

Figure 2.5 shows a flow diagram of an open-loop system. The basis of open-loop systems is that the system is controlled by a signal which is at a pre-set value. This pre-set value assumes the required control can be achieved without measuring the effect of the system output on the parameter it is set to control. The pre-set value will not change even if other factors do which render the system output incorrect.

Consider an open-loop system for switching on and off the street lamps along a road. The control required is that when darkness falls, the lights turn on, and when it is light, the lights turn off. The control signal would be pre-set to accordingly switch the lights on or off at set times in the evening and morning, using a timing device. This system would probably work acceptably for a few weeks. However, because the hours of darkness change over the year, the pre-set signal would soon be incorrect and the lights would be switching on and off too early or too late. Figure 2.6 shows this open-loop street lighting system in flow diagram form.

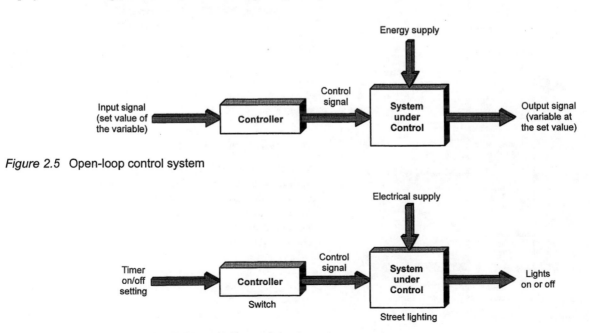

Figure 2.5 Open-loop control system

Figure 2.6 Timer based street lighting system

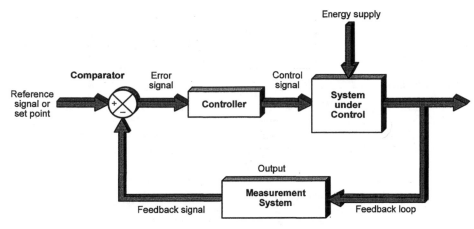

Figure 2.7 Closed-loop control system

There is no input to the open-loop system which detects what is actually happening with the parameter the system is affecting, that is, the street lamp system doesn't know whether it is light or dark. As an open-loop system, someone has to estimate when it gets dark and light and set the timer control which turns the lights on and off accordingly. These pre-set times have to be altered as the hours of light and dark vary throughout the year. Hence it requires frequent operator intervention, because the less often these settings are updated, the less efficient they will be. The system will also not take into account any unexpected or unpredictable behaviour. For example, some evenings may be dull and cloudy, so ideally the lights would turn on earlier than on a bright, sunny evening.

Open-loop control systems are in general relatively simple in design and inexpensive. However, they can be inefficient and require frequent operator intervention. Under many circumstances pre-set values become incorrect due to the parameter they are controlling changing in some way, and so they need resetting. The pre-set value often needs a high level of skill or judgement to set it correctly. In cases where the consequences of the system not controlling the parameter as desired are serious, such as the level of a hazardous liquid in a tank overflowing, open-loop systems are not suitable.

Closed-loop control systems

In a closed-loop control system, the state of the output directly affects the input condition. A closed-loop control system measures the value of the parameter being controlled at the output of the system and compares this to the desired value.

Key fact

In a closed-loop control system, the actual value of the parameter being controlled is compared to the desired value. The difference in these values is known as the error.

Referring to Figure 2.7, which shows a closed-loop system in flow diagram form, the desired value is known as the reference signal, or set point. This is compared to the signal from the measurement device, known as the feedback signal. The difference between the feedback signal and the reference signal is known as the error signal. The error signal is then modified so that it can adjust the performance of the system. For example, if the error signal is an electrical signal, it may need amplification. The modified error signal is called the control signal. The control

signal then adjusts the output of the system, to try to match the feedback signal to the reference signal. This would reduce the error signal to zero and so achieve the desired value.

Consider a tank in a chemical plant, containing a hazardous liquid, as shown in Figure 2.8. The tank is filled with liquid by a pump. When the liquid is required for further chemical processing, another system opens the discharge valve and takes the liquid it needs, reducing the tank level. It would not be acceptable to use an open-loop system because the consequences of a pre-set value becoming incorrect would be serious. The tank may overfill, spilling hazardous liquid, or run dry and stop operation of the plant. For the plant to work effectively the tank needs to be filled to an optimum level. A liquid level sensor detects the level and produces an electrical output (we shall discuss sensors like this in Chapter 4).

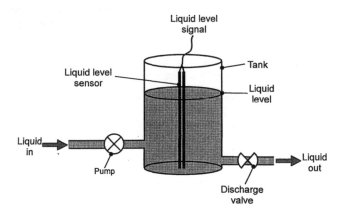

Figure 2.8 Control of liquid level in a tank

Figure 2.9 shows a flow diagram of closed-loop control of this system. The output from the liquid level sensor, the feedback signal, is compared to the ideal level as dictated by the reference signal. This produces the error signal. The error signal is amplified by the controller to become the control signal. The control signal drives the pump motor and so determines the flow rate of liquid through the pump to fill the tank. When the error signal is zero, the liquid level has reached its ideal level, the control signal is also zero and so the pump stops. In this way the information contained in the electrical signal concerning the liquid level, whether it be constant or varying, controls the pump flow rate to maintain the liquid level under varying discharge conditions.

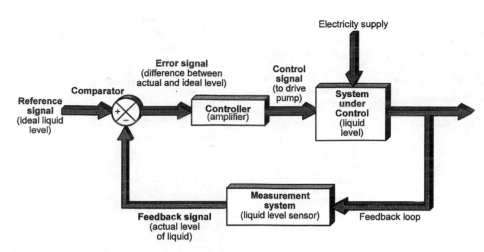

Figure 2.9 Flow diagram of closed-loop control of liquid level in a tank

Closed-loop systems regulate themselves and therefore are less prone to error than open-loop systems. They are generally more efficient and require less operator involvement. However, set up costs can be high and they can become complex.

Sensor terminology

Some of the terminology used with sensor measurement and control systems may be unfamiliar to you. Some of the definitions may have different meanings in other fields, and to use this book effectively it is essential that you have a basic understanding of sensor terminology. Most of the terms used are explained in the book as we meet them, for example we have already discussed closed-loop control systems. However, we need to define a few concepts here before continuing with other chapters. Parts of this book also require a basic understanding of a.c. and d.c. theory.

Absolute measurement

Absolute measurement uses measurement scales based on the fundamental units of a system. It relates to the condition at which a system contains none of the variables to be measured. This is opposed to arbitrary scales which relate to a predefined numerical value. For example, when discussing temperature in Chapter 6 we will meet the Kelvin scale and the Celsius scale. The Kelvin scale is absolute because it relates to the condition where a substance has no temperature, but the Celsius scale is arbitrary because its zero point relates to a predefined value of temperature (the melting point of water).

Conditioned signal

A conditioned signal is the output from a sensor which has been modified so it can be understood by a display or recording device, a control device, or other system.

Control

In terms of measurement systems, control means to operate or regulate a machine or system.

E.m.f.

E.m.f. stands for electromotive force. This is a source of energy in an electrical device or circuit which can cause a current to flow. It is the rate at which this energy is drawn from the device when current flows. E.m.f. is measured in volts.

Fluid

By definition a fluid is any substance which flows. Fluids are usually gases or liquids, but are sometimes a collection of solids such as powders or sand. They have no fixed shape and offer little resistance to stress.

Flux

In later chapters we refer to magnetic flux and electric flux. Here, the flux refers to the strength of the electric or magnetic field in a given area.

Integrated circuit

An integrated circuit is an electronic circuit made from a wafer of semiconductor material, which cannot be separated into its individual components.

Interfacing

Linking two electronic devices by designing or adapting their inputs and outputs so they work together compatibly.

Local reading

A sensor which is read locally displays its measured values at, or very close to, the point of measurement.

Measurand

The measurand is the input to the measurement system, the quantity or parameter which is to be measured. For example, a thermometer measures temperature, therefore the measurand of a thermometer is temperature. The measurands covered in this book include displacement, velocity, proximity, acceleration, level, height, weight, volume, fluid flow, temperature and pressure.

Electrical noise

Electrical noise is the presence of unwanted electrical signals. These can obscure or confuse the signal which carries useful information, such as the sensor output signal or the error signal.

Parameter

A parameter is a variable quantity with defined limits.

Probe

A probe is a lead or device which connects from a sensor or display to the measurand or an electrical circuit.

Remote reading

A sensor which can be read remotely can display its measured values away from the point of measurement.

Semiconductor

Semiconductors are substances having electrical properties somewhere between metals and insulators. Devices made from semiconductors form the basis of modern integrated circuits and computers.

SI units

SI units are fundamental units of measurement used by international agreement, to ensure scientific and technical consistency. The 'Système International d'Unités' (International System of Units, or SI) uses whole, multiples, or divisions of these units. The units we will use in this book can all be derived from the basic SI units of the metre, kilogram, second, ampere, kelvin, and radian.

Specification

The specification of a device is a technical description of its characteristics, construction, performance and any other information relevant to its use. Several sample specifications illustrating typical sensor characteristics are given in this book.

Variable

In the context of sensors and measurement systems, a variable can be considered as anything, usually a quantity or measurand, that can change in value.

Sensor characteristics

Choosing a sensor for a measurement or control system depends on several factors, such as cost, availability, and environmental factors. When choosing a sensor it is important to match its characteristics to the quality of output required. There are, for example, many types of sensor available to measure temperature, but not all would be suitable for displaying the air temperature to a gardener in the example we discussed earlier. Some would not be able to measure the range of temperature, some would be too expensive, or require an electrical supply.

As with the terminology defined earlier, it is essential that you have a basic understanding of sensor characteristics to be able to use this book. Some definitions of sensor characteristics may be familiar to you, as many of the words are used in everyday language. However, consider the following definitions carefully, as they may be slightly different when applied to sensors and measurement systems than in conventional use.

The characteristics which follow can be applied to the whole measurement system and all components within a measurement system, including the sensor, signal conditioning unit, and display or recording device. There are different ways of expressing all of them, but usually they are stated as a percentage or as a maximum and minimum value, depending on the nature of the system and measurand, and the manufacturers preference. Not all the characteristics will necessarily apply to a given sensor.

Accuracy

The accuracy of a device or system is the extent to which any value it creates could be wrong, or the maximum error it may produce. With a sensor, it is how close the output value is to the actual value of the measurand. In practice, every device will produce an error, however small, and will have some degree of accuracy rating. It may be expressed in terms of the measurement units involved, for example, suppose we have a thermometer with an accuracy of ±0.2 °C. This means that if we take a temperature measurement with the thermometer and find it to be 20.1 °C, then the actual temperature lies somewhere between 19.9 °C and 20.3 °C. Alternatively it may be expressed as a percentage error of the range of the device.

Calibration

Calibration refers to the units the scale of a sensor display or recorder is labelled in. For example, a type of sensor measuring the speed of a vehicle produces an electrical output. The size of this voltage is proportional to the speed of the vehicle. The speedometer pointer in the vehicle moves with respect to the voltage applied to it, but will be labelled in units of speed, not voltage. We therefore say the speedometer is calibrated in terms of speed.

Dead-zone or dead-band

When a specification refers to a dead-zone or dead-band, it is referring to the largest change in the quantity to be measured to which the output does not change, or the range of input to which there is no output. Dead-zones arise as a result of static friction (stiction), or hysteresis (explained later in this section). Figure 2.10 is a graph showing dead-zone characteristics.

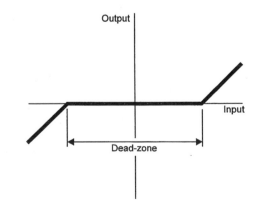

Figure 2.10 Dead-zone or dead-band

The dead-zone may not apply over the full operating range of the device, and sometimes significant dead-zones only occur under certain conditions. A common example of a dead-zone is a dimmer switch on a domestic light or lamp. Often, when the dimmer switch is turned fully down, it can be turned up slowly and for a short while there will be no immediate response from the light. In this case the dead-zone of the dimmer switch is from its fully turned down position to the position it is in when the light first starts to glow.

Dimensions

The dimensions of a sensor or measuring system are a measurement of its physical size, and are shown on nearly every device specification.

Drift

Drift is the natural tendency of a device, circuit, or system to alter its characteristics with time and environmental changes. There is a change in the output characteristics while the inputs to the device have not changed, which affects accuracy. Drift occurs over different time scales for different reasons. One of the most common and influential drift inducing effects is a change in ambient temperature. This is why the specifications of many sensors state the effect of temperature on various characteristics of the device. On an older device drift may be caused by ageing affects on the materials it is made of, such as oxidation of metal elements. It can also be caused by the mechanical wear or self-heating of components in a system.

Error

Error is the difference between a measured value and the actual value. For example, a ruler is used to measure the width of a page of a book, and it is found to be, say, 210.5 mm. However, the actual size of the page measured was 209.9 mm, therefore there was an error in measuring the size of the book of 210.5 mm − 209.9 mm = 0.6 mm. Error may often be quoted as a percentage to represent the accuracy of a system.

Hysteresis

Hysteresis causes the difference in the output of a sensor when the direction of the input has been reversed. This produces error and so affects the accuracy of a device.

Figure 2.11 shows this in graphical form. The input to the sensor, the measurand, is increased in set increments. Towards the end of its range, the measurand is decreased in similar sized decrements. The graph shows the difference in the output of the sensor when the measurand value is increasing to when it is decreasing. This is the hysteresis of the system.

Not all sensors or measurement systems suffer from hysteresis. It is caused by various factors, particularly mechanical strain and friction. Slack motion in gear systems and screw threads (often referred to as backlash) is also a common cause. Hence measurement systems likely to suffer significantly from hysteresis may incorporate mechanical gears or bearings, or other moving parts, and materials which tend to be elastic, such as rubber, plastics, and some metals.

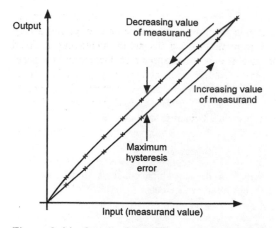

Figure 2.11 Graph of the effect of hysteresis

Lag

Lag is the delay in the change of the output from a sensor with respect to a corresponding change in the input. It is measured in seconds (or more usually fractions of a second). In some applications, such as control, lag can seriously affect performance.

Linearity

The linearity of a sensor is the amount by which the graph of its input against output is near to being a straight line. It may be quoted as being linear for a range of input values, as shown in Figure 2.12. It may also refer to the maximum amount by which the graph deviates from being a straight line, quoted as a percentage of the operating range.

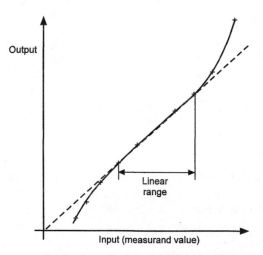

Figure 2.12 Linearity

Operating life

The useful operating life of a sensor is an indication of how long it can be expected to function within its specification. It is expressed in terms of time, or in the number of operations or cycles it should successfully endure.

Precision

The precision of a device is the degree to which it produces similar results for the same input on a number of occasions.

Precision is often used in everyday language to mean accuracy, and hence the two are sometimes confused. However, in measurement terminology a sensor can be precise, giving a very similar output a number of times when measuring a fixed quantity, but, if there is a large error in the output it is not accurate. On a specification, precision is usually quoted in general terms (such as a high precision instrument). Repeatability (reproducibility) is a quantified measure of precision, and more likely to be quoted on a specification.

It is important to note there are varying definitions of precision. In this book we will use its meaning as defined above. However, in some specifications or other texts it may refer to resolution.

Range

The operating range of a device is a statement of the limits in which it can function effectively. The operating range of a sensor is usually specified as the lowest and highest input values which it is capable of measuring. Other ranges are frequently quoted on device specifications, for example temperature range refers to the maximum and minimum temperatures in which the device will work. Humidity and pressure ranges are also commonly quoted. It is important not to exceed these ranges because not only may the device fail to operate effectively, it could also damage other components of the system or even be a risk to health and safety.

Rating

The rating of a device relates to the recommended conditions, electrical or mechanical, under which it will successfully or safely operate. A description of the type of rating is usually given, for example maximum temperature rating, or average load rating.

Reliability

The reliability of a device is similar to its operating life, and may often be quoted instead, depending on its nature. Reliability is the ability of a device to perform its function under specified conditions for a stated period of time, or a stated number of operations, whilst remaining within its specification.

Repeatability

Repeatability is a numerical measurement of precision under fixed conditions. It is a measurement of the ability of a device to produce identical indications or responses, for repeated applications of the same value of the physical quantity to be measured. It may be given as ± a maximum percentage of the reading, or within stated limits of each reading.

Reproducibility

Reproducibility is another term for repeatability.

Response

The response of a device is the time it takes to reach its final output value for a given input. It may be quoted in terms of seconds or fractions of a second, or sometimes as a percentage of its full value. For example, if a specification states that 95% response time is 3 seconds, it means that the device takes 3 seconds to reach 95% of its final output value.

Resolution

The resolution with which a device senses or displays a value relates to the smallest input or change in input it is able to detect. It is usually expressed in terms of the smallest increment which can be measured or sensed. The higher the resolution of a display, the smaller the increment it is able to measure. For example, a five-digit display which can measure a quantity to 0.0001 units has a higher resolution than a four-digit display measuring to 0.001 units. It is usually expressed as a percentage.

Sensitivity

Sensitivity is the relationship under fixed conditions between a change in the output of a device to the change in input. The sensitivity of a sensor is the difference in its output values over a given range divided by the change in the value of the measurand. That is,

$$\text{Sensitivity} = \frac{\text{Maximum output value} - \text{minimum output value}}{\text{Maximum input value} - \text{minimum input value}}$$

The units in which sensitivity is expressed are defined by the above equation and consequently vary depending on the nature of the device and measurand. For example, as we shall see later there are sensors which measure small distances moved by an object in terms of voltage. In this case the units of sensitivity would be volts per millimetre.

If the relationship between the measurand and the output is linear, then the sensitivity will usually be expressed over this whole range. If it is non-linear, the sensitivity characteristics of the device will vary for different values, and so the sensitivity will usually be quoted for several ranges.

Stability

Stability is a measurement of how much the output from a device or system varies if, under fixed conditions, a constant input is applied over a long time.

Static error

Static error is a constant error that occurs throughout the input range of the device. If static error is known it can be compensated for and not significantly affect accuracy.

Tolerance

The tolerance of a device is the largest amount of error that can occur during its operation. Depending on the nature of the device, tolerance may sometimes be quoted instead of accuracy.

Summary

This chapter has introduced you to sensors and sensor technology, and discussed the measurement and control systems in which they are used. It has also defined some basic terminology and characteristics used when discussing sensors, measurement, and control systems.

You will meet many more definitions and concepts in the following chapters, but in the course of reading this book it may at times be useful to refer back to this chapter.

The following chapters look in detail at a selection of different types of sensors. Sensors and sensing instruments are grouped together according to their measurands, such as motion, level, height, weight, volume, fluid flow, temperature, and pressure.

Questions for further discussion

1. Consider a dynamo on a bicycle, which converts some of the kinetic energy from the rotation of one of the bicycle wheels into electrical energy. Is this a sensor?

2. Consider the timer controlled street lighting system discussed earlier. Still keeping the system as open-loop, how could it be improved? What potential problems could you foresee with the closed-loop control we discussed?

3. What do you consider to be the main differences between the sensor characteristics of rating and range? Under what circumstances could one be quoted instead of the other, and similarly under what circumstances could they both be quoted on the same specification?

4. List the characteristics which could be included in the specifications of (a) a thermometer for use by a gardener (b) a set of kitchen scales, and (c) a speed indicator for a bicycle. Where appropriate, indicate which units (or percentages) would be most suitable to express each characteristic.

3 Measurement of Motion

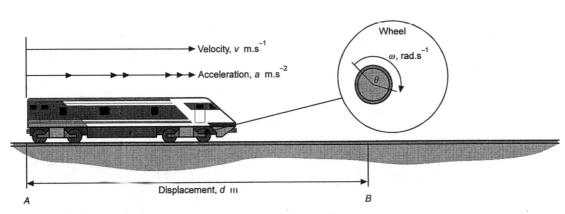

Figure 3.1 Linear and angular motion

In this chapter we shall consider sensors and transducers used to measure motion. Motion is the change in the physical position of an object, and encompasses several variables. For example, consider a train travelling along a track, as shown in Figure 3.1.

The vehicle is moving in a straight line through a displacement d metres between points A and B. At any point in time it will be moving at a linear velocity v metres per second in the direction indicated by the arrow, and it will be accelerating or decelerating at a rate a metres per second per second. If we look at a wheel of the train, not only does this have linear motion but, because it is rotating, it also has angular motion. If it turns through an angle θ radians, it will have travelled at an angular velocity ω radians per second, and if it had been accelerating it would have accelerated at a rate α radians per second per second.

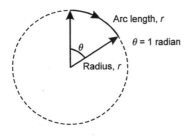

Figure 3.2 The radian

Notice the units used to measure motion. For linear motion, the fundamental unit of length is the metre. By international convention this is currently defined as the length of the path travelled by light in a vacuum during a time interval of 1/299792458 of a second. For angular motion, the basic unit is the radian. A radian is defined as the angle subtended at the centre of a circle by an arc equal in length to the radius. Figure 3.2 shows this in graphical form. There are 2π radians in one revolution or

360° (because the circumference of a circle is $2\pi \times$ radius), while one radian is equal to about 57.3°, and 1° is equal to about 0.0175 radians.

In this chapter we shall consider sensors which measure the following types of motion:

- Linear displacement.
- Angular displacement.
- Proximity.
- Acceleration.

Measurement of displacement is important because many systems have an input or output which is in the form of a displacement. However, the displacement they measure may relate to, and so be expressed in terms of, another parameter. For example, a strain gauge load cell (which we shall meet in Chapter 4) measuring force is actually measuring linear displacements caused by force, using strain gauges, and expressing these displacements in terms of force. Similarly a spring balance indicating force is actually measuring the displacement of a spring. The linear speed of a car as indicated by its speedometer is measured in terms of the rotational speed of the final drive from the gear box. A dial gauge, such as on a pressure gauge, indicates pressure in terms of a calibrated rotary displacement.

Many linear and angular displacement devices are capable of measuring displacement against time, and therefore velocity and acceleration. Proximity sensors sense the presence of an object. If the presence of the object is sensed at regular intervals, for example the teeth of a gear wheel, it allows the velocity and acceleration to be determined. Other devices directly measure velocity, or acceleration (accelerometers), from which it is possible to determine displacement.

Linear displacement

Displacement is the distance measured in a stated direction from a reference point. If it is measured in a straight line it is called linear displacement, and if it is measured in terms of an angle of rotation about an axis it is called angular displacement.

> **Key fact**
>
> Displacement is the magnitude and direction representing a change in position of a body or point with respect to a reference point. Linear displacement is a displacement in a fixed direction. Angular displacement is the angle turned through by a body about a given axis.

Large displacements requiring highly accurate measurements are usually measured by radar, sonar, or sophisticated optical systems, which are beyond the scope of this book. The devices we'll consider here generally measure linear displacements of less than one metre.

There are many simple mechanical devices available for measuring linear displacement These include rulers, slip gauges, micrometers and vernier gauges. They are quick and easy to use, but their accuracy relies heavily on the skill of the operator. They cannot usually be used for remote measurement.

A mechanical linear displacement measuring device which is generally more accurate than other mechanical devices, and also some electrical devices, is the dial test indicator.

Dial test indicator

Mechanical dial test indicators are commonly used in a wide range of applications for measuring displacement, and may often be referred to as DTIs.

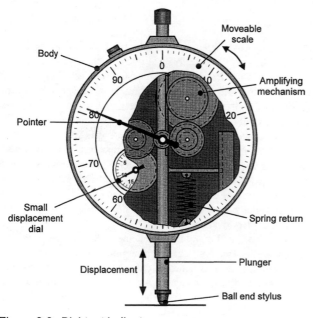

Figure 3.3 Dial test indicator

Figure 3.3 shows a typical mechanical dial test indicator. The body of the indicator is fixed in relation to the object whose displacement is to be measured. A spring-loaded plunger fitted with a stylus touches the object. The stylus can be one of a number of designs, such as ball-ended or roller ended. The

displacement is amplified by means of gears and displayed on a dial. A moveable scale on the dial can be rotated to obtain an appropriate reference point. The plunger only records the displacement of the object in one plane.

To improve the accuracy of reading, there are usually two dials. The large dial shows displacement in small increments (typically 0.01 mm), with the smaller dial displaying the same displacement but in larger units (typically 1 mm). A digital display is used instead of a mechanical dial gauge in some types of DTI.

Dial test indicators are quick and easy to use, and are accurate. They are widely used for measuring displacement in manufacturing and throughout the engineering industry. Typical applications are checking the dimensions of products for quality control, setting machines correctly, and checking components for wear.

Dial test indicators usually have to be read locally and cannot easily measure displacements which change in direction. The displacement they are measuring has to be accessible to the plunger, and able to withstand the spring force of the plunger (even though this is usually small). Therefore parts that are fully enclosed, obscured by other items, or delicate may not be suited to this device.

Linear potentiometer

Potentiometers are electrical devices which are a form of variable resistance. Figure 3.4 shows a typical layout of a linear potentiometer. It consists of a sliding contact which moves over the length of a resistance element. This sliding contact connects to a plunger, which links to the object whose displacement is to be measured.

There are several designs of sliding contact (sometimes called wipers), depending on their application. Sliding contacts are often made of copper alloys. This is because copper alloys have elastic properties so can be formed to maintain good electrical contact with the resistance element. They are also good electrical conductors. Resistance elements are commonly made of thin nickel or platinum wire wound around a former made of insulating material. Resistance elements may also be made of films of carbon, metal, or conductive plastics to improve resolution. The linear potentiometer shown in Figure 3.4 has two sliding contacts and a guide. These keep the movement of the sliding contact over the resistance element smooth and constant.

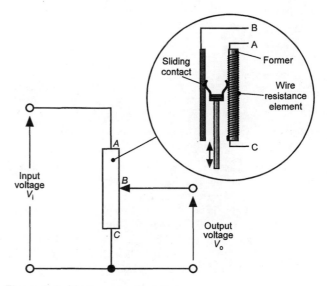

Figure 3.4 Linear potentiometer

Problem

Consider the linear displacement potentiometer circuit diagram shown. The input voltage V_i is 5 volts, and the output voltage V_o is 2.5 volts. The total resistance element length is 100 mm, so when the sliding contact is at the centre, the distance $AB = BC = 50$ mm.

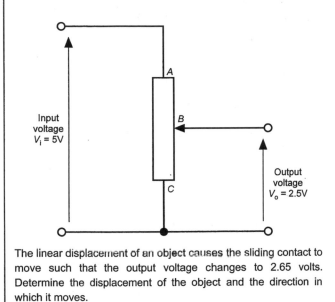

The linear displacement of an object causes the sliding contact to move such that the output voltage changes to 2.65 volts. Determine the displacement of the object and the direction in which it moves.

Solution

Voltage across AC = 5 volts.

Distance AC = 100 mm

$$\therefore \text{Volts mm}^{-1} = \frac{5}{100} = 0.05 \text{ V.mm}^{-1}.$$

If the output voltage changes from 2.5 volts to 2.65 volts, this is a change of 2.65 − 2.5 = 0.15 volts.

Thus the displacement of the object = $\frac{0.15}{0.05}$ = 3 mm

Therefore the displacement = 3 mm. Since the output voltage has become more positive, the displacement is towards A.

Figure 3.5 Sample calculation of displacement with a linear displacement transducer

Referring to the electrical circuit shown in Figure 3.4, an input voltage V_i is applied across the whole resistance element, at points A and C. The output voltage, V_o, is measured between the sliding contact at point B and the end of the resistance element at point C. A linear relationship exists between the input voltage V_i, output voltage V_o and the distance BC.

To measure the displacement of an object, as the object moves, the plunger moves and transfers this displacement to the sliding contact. Hence any displacement of the object will alter the distance BC, causing a corresponding change in the output voltage V_o. This output, V_o, which represents the displacement of the plunger, is shown on a voltmeter, calibrated in appropriate units.

Figure 3.5 gives an example calculation for finding the displacement of an object from the output voltage of a linear potentiometer.

Potentiometers suffer from some non-linearity of the former, which affects the accuracy of the results. The sliding contacts and the resistance elements are often prone to wear, which adversely affects their performance. They also add some physical resistance to displacement. Problems may also be caused by unwanted electrical signals (electrical noise).

Wire-wound potentiometers have a linearity of about ±1%, and the more expensive film types can be linear to within 0.01%. The resistance of wire-wound potentiometers ranges from about 10 Ω to 200 kΩ, and for film types from about 100 Ω to 1 MΩ. The resolution of wire-wound potentiometers depends on the number of windings of the resistance element.

Direct indication can be achieved using a voltmeter calibrated in units of distance. Remote reading or recording can be achieved by using the voltage change across the output as an input signal to a measuring or recording system. Linear potentiometers are often used where an electrical signal relating to displacement is required, but where costs should be kept low and high accuracy is

not paramount. They are used in a variety of applications, for example position monitoring of products on assembly lines, for checking dimensions of products in quality control systems.

There are several different types of potentiometer apart from linear potentiometer, which have different output characteristics. These include the logarithmic potentiometer, the cosine potentiometer, and the rotary potentiometer. We shall discuss the rotary potentiometer later in this chapter.

Linear variable differential transformer (LVDT)

Linear variable differential transformers (usually referred to in their abbreviated form of LVDT), are probably the most commonly used sensors for accurately measuring displacements up to about 300 mm.

A conventional transformer consists of two closely coupled coils wound around a soft iron former. These are known as the primary coil and the secondary coils. When an a.c. voltage is applied to the primary coil, an a.c. voltage is induced in the secondary coil. This is because of electromagnetic induction.

Key fact

Faraday's Law of Electromagnetic Induction states that when a conductor moves through a magnetic field an electromotive force (e.m.f.) is induced across it proportional to the rate of cutting flux.

When an alternating current (a.c.) flows through the primary coil it produces an alternating magnetic flux. Because of Faraday's Law of Electromagnetic Induction, an e.m.f. is induced in the secondary coil. The size of the e.m.f. induced in the secondary coil depends on the amount of the current flowing in the primary coil and the ratio of primary and secondary turns. Hence the voltage across the secondary coil depends on the ratio of turns in the windings.

The linear variable differential transformer is a precision instrument designed and used for measuring displacement.

The linear variable differential transformer is aptly named since its operating principles can be readily obtained by considering its name word by word, in reverse order.

Firstly, it is a transformer, obeying all of the principles of electromagnetic induction appropriate to this type of device.

It has one primary winding and two secondary windings connected to provide the difference in their respective voltage levels at the output. This is the differential.

It is variable because the magnetic coupling between the primary and each of the two secondary coils can be varied to affect the magnitude of the induced e.m.f.s.

Finally, the design of the whole assembly is such that the variation in the coupling between the primary and the secondary coils is linear.

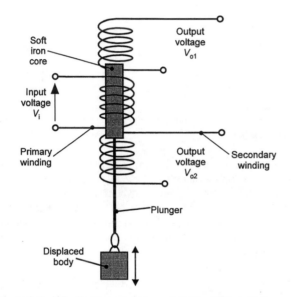

Figure 3.6 Windings on a linear variable differential transformer (LVDT)

Figure 3.6 shows the relative position of the three windings of an LVDT wound on a hollow former. They lie along a single axis, the primary winding in the centre and the two secondary windings on either side. A soft iron core is positioned in the centre of the windings which is free to move in either direction within the coils.

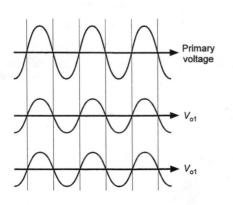

Figure 3.7 Induced e.m.f.s in LVDT secondary windings

Initially assume that the soft iron core is centrally positioned within the primary winding. When energised by an a.c. signal, usually at a frequency of 5 kHz or above, the current flowing produces a magnetic flux in the soft iron core. This flux links with both of the secondary windings equally and these have identical e.m.f.s induced across them. This is shown in Figure 3.7.

The secondary voltages, V_{o1} and V_{o2} are in phase with each other and have the same amplitude.

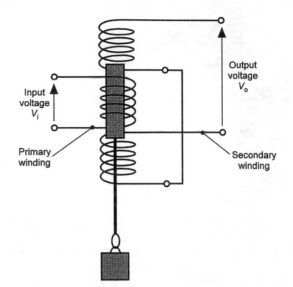

Figure 3.8 Linear variable differential transformer (LVDT)

If the two secondary windings are now connected as shown in Figure 3.8 then the two signals will cancel.

However, should the soft iron core be moved in either direction the flux linkage to one secondary winding will increase while the flux linkage to the other secondary winding will reduce. Similarly, should the core be moved in the opposite direction the effect will be reversed. Figure 3.9 shows the relative amplitude of the combined outputs from the secondary windings against displacement in either direction.

Note that with no displacement the secondary voltage is zero. This voltage increases with displacement in either direction. Eventually the linkage is wholly with one secondary winding and not the other and so the output voltage is at a maximum and unable to increase further (saturated). This limits the effective working range of the LVDT.

For displacements in one direction only, positive or negative, a measure of the secondary voltage amplitude alone would give an indication of displacement.

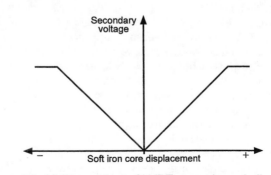

Figure 3.9 Voltage output of LVDT secondary windings against displacement

Where displacement is expected to be in either direction then further signal conditioning is required. This can be achieved by comparing the secondary output voltage with a reference, usually the primary supply. Whichever of the two individual secondary voltages is largest controls the phase of the combined secondary output, either in-phase or antiphase (assuming no inductive effects in the windings causing their own phase shifts). A phase sensitive detector may be used to compare the secondary output voltage with the reference. This produces a signal proportional to the amplitude of the secondary with a polarity governed by whether it is in phase or anti-phase with the reference. To take advantage of this effect to measure displacement only requires the LVDT soft iron core to be connected to the object.

Some commercially available LVDTs can be supplied from a d.c. source, and provide a d.c output. They are based on the same principles as a.c. LVDTs but have built in signal conditioning. The d.c. is converted to a.c. before being input to the LVDT, and the output is then converted back from a.c. to d.c. This gives a d.c. output dependent on the position of the soft iron core.

The LVDT is extremely sensitive and provides resolution down to about 0.05 mm. They have operating ranges from about ±0.1 mm to ±300 mm. Accuracy is ±0.5% of full-scale reading. Because there is no contact between the magnetic core and the coils, there is very little friction and wear. If necessary they can be constructed to withstand shock and vibration. Consequently LVDTs are in wide use in various applications. These range from use in machine tools, to robotics and digital positioning systems.

LVDTs often form part of systems which measure force, pressure and acceleration.

The bonded resistance strain gauge and the Wheatstone bridge

Bonded resistance strain gauges measure changes in the size of a solid object, due to the object being strained (stretched or compressed). They are bonded to the object so that as the object changes size, so do they.

Although there are other types of strain gauge available, bonded resistance strain gauges are probably the most commonly used. For convenience they are usually referred to in the shortened form, simply as strain gauges.

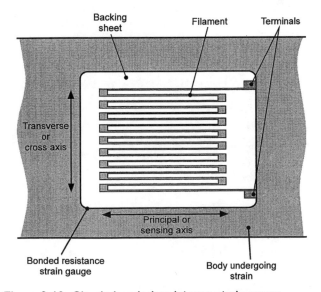

Figure 3.10 Simple bonded resistance strain gauge

Strain gauges are sensors that experience a change in electrical properties when their dimensions change. When resistance strain gauges stretch or compress, their resistance changes, and this change in resistance can be shown in terms of displacement. Bonded resistance strain gauges can measure strains over lengths up to about 50 mm, the total displacement being a small percentage of this (usually around 1%).

Figure 3.10 shows a simple strain gauge. It consists of a metal foil mounted on a backing sheet. The foil is printed, or etched using photographic techniques so it is in the form of a continuous strip in a zigzag pattern. This is called the filament. Strain gauge filaments are made of materials such as copper-nickel or nickel-chromium alloys which have high resistivity and mechanical strength. They are usually only a few micrometres thick. The backing sheets are made of various types of epoxy resin, the specific choice depending on the application, and are also very thin. Sometimes the whole gauge may be encapsulated.

Strain gauges vary in size from fractions of a millimetre up to around fifty millimetres. They are usually fixed to the object by special adhesive cements, but may sometimes be embedded or welded. The adhesive used to bond the strain gauge must be strong and an electrical insulator.

When a strain gauge is bonded to an object, and the object (and so the strain gauge) changes in size, the resistance of the strain gauge filament will change. The resistance of the filament R is given by the expression:

$$R = \frac{\rho l}{A}$$

where:

- ρ is the resistivity of the material in ohm meters
- l is the length of the filament in metres
- A is the cross-sectional area of the filament in metres squared

From the expression we can see that changing either the area of the filament A, or length of the filament l (or both), will change the resistance R. Stretching the filament causes its length l to increase and its cross-sectional area A to decrease. Therefore stretching will change its electrical resistance R. The strain gauge uses this effect to measure displacement.

A linear strain gauge is sensitive to changes in length along the principle or sensing axis, and virtually insensitive to changes in length along the transverse or cross axis. Consequently, sound bonding and correct positioning of the strain gauge on the object whose dimensions are being examined is essential to accuracy.

When strain gauges measure the changing dimensions of an object, they are measuring strain (and hence their name). Strain is the ratio of the change in dimension of an object to the original dimension. Because it is a ratio it has no units.

$$\text{Mechanical strain } \varepsilon = \frac{\text{Change in length}}{\text{Original length}} = \frac{\Delta l}{l}$$

$$\text{Electrical strain } G\varepsilon = \frac{\text{Change in resistance}}{\text{Original resistance}} = \frac{\Delta R}{R}$$

where G is the gauge factor, defined as the ratio of the fractional resistance change to the applied strain, that is

$$G = \frac{\left(\frac{\Delta R}{R}\right)}{\left(\frac{\Delta l}{l}\right)} = \frac{\left(\frac{\Delta R}{R}\right)}{\varepsilon}$$

$$\therefore G\varepsilon = \frac{\Delta R}{R}$$

The gauge factor G of strain gauges usually lies in the range 1.8 and 2.2. Resistances of strain gauges are available in the range 50 Ω to 2 kΩ.

We use the unit of microstrain, $\mu\varepsilon$ to describe the ratio ε. To ensure a linear relationship between ΔR and change in dimension, the value of ε must be small. For example, if the length of an object l is 100 mm, and it increases in length Δl by 0.1 mm, then the strain ε will be:

$$\varepsilon = \frac{\Delta l}{l} = \frac{0.1}{100} = 0.001$$

This would normally be written as 0.001×10^6 $\mu\varepsilon$ or 1000 $\mu\varepsilon$. Figure 3.11 is an example calculation determining the displacement of an object using a strain gauge. We have already seen that a strain gauge exhibits a change in resistance when it is strained. This change in resistance needs to be converted into an electrical signal which can be used to show the straining force or displacement of the object. A circuit commonly used to enable this is the Wheatstone bridge.

A Wheatstone bridge is an electrical circuit for determining resistance. A Wheatstone bridge circuit is said to be balanced when the resistances in the circuit are adjusted so there is no voltage across the output.

Referring to Figure 3.12, the output voltage V_o across BD is zero when the bridge is balanced. If the bridge becomes unbalanced V_o will move away from zero and its polarity will depend on which of the resistors has changed and whether it has increased or decreased. If R_1 is replaced by a strain gauge and R_2,

R_3 and R_4 are held constant, changes in resistance due to strain or displacement will unbalance the bridge. The value of V_o is a measure of the applied strain and this change can be used as an input to a measuring or control system.

We shall look more closely at Wheatstone bridges and the relationship between the change in output voltage and the change in strain gauge resistance in Chapter 9. We shall find that this relationship can be stated as:

$$\Delta V_0 = V_i\left(\frac{\Delta R_1}{R_1 + R_2}\right)$$

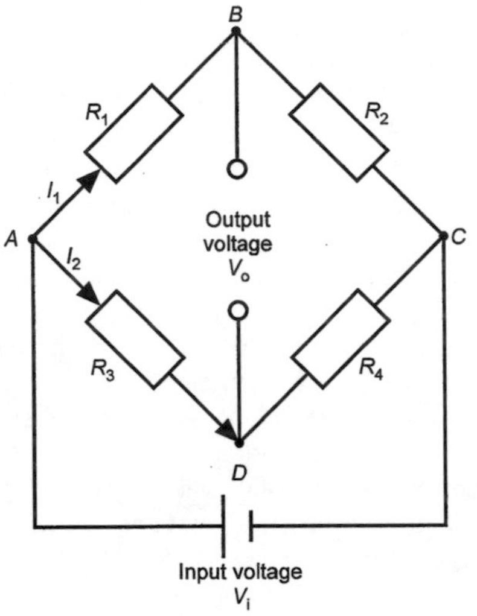

Figure 3.12 Wheatstone bridge circuit

Problem
A strain gauge has a resistance of 250 Ω and a gauge factor of 2.2. It is bonded to an object to detect movement. Determine the change in resistance of the strain gauge if it experiences a tensile strain of 450 $\mu\varepsilon$ due to the change in size of the object. Also, if the relationship between change in resistance and displacement is 0.05 $\Omega.mm^{-1}$, determine the change in size of the object.

Solution
From the question, we know that R = 250 Ω, G = 2.2, and $\varepsilon = 450 \times 10^{-6}$.

$$G\varepsilon = \frac{\Delta R}{R}$$

$$2.2 \times 450 \times 10^{-6} = \frac{\Delta R}{250}$$

$$\Delta R = 250 \times 2.2 \times 450 \times 10^{-6}$$

$$= 0.2475\ \Omega$$

Therefore, the change in resistance of the strain gauge, $\Delta R = \mathbf{0.2475\ \Omega}$

To find the displacement Δl,

We know from the question that

$$\frac{\Delta R}{\Delta l} = 0.05\ \Omega.mm^{-1}$$

$$\therefore \Delta l = \frac{\Delta R}{0.05\ \Omega.mm^{-1}}$$

$$\Delta l = \frac{0.2475\ \Omega}{0.05\ \Omega.mm} = 4.95\ mm$$

The change in size of the object, $\Delta l = \mathbf{4.95\ mm}$

Figure 3.11 Example calculation determining the change in size of an object using a strain gauge

Problem

A strain gauge bridge has a strain gauge of resistance $R = 200\ \Omega$ and gauge factor $G = 1.9$. R_2, R_3 and R_4 are fixed resistors also rated at $200\ \Omega$. The strain gauge experiences a tensile strain of 400 microstrain due to the displacement of an object. Determine the change in resistance ΔR of the strain gauge. If the input voltage V_i is 4 volts then determine the change in output voltage ΔV_o.

Solution

We know that $R = 200\ \Omega$, $R_2 = 200\ \Omega$, $R_3 = 200\ \Omega$, $R_4 = 200\ \Omega$, $G = 1.9$, and $\varepsilon = 400 \times 10^{-6}$. At balance $R = R_2 = R_3 = R_4 = 200\ \Omega$.

$$G\varepsilon = \frac{\Delta R}{R}$$

$$1.9 \times 400 \times 10^{-6} = \frac{\Delta R}{200}$$

$$\therefore \Delta R = 200 \times 1.9 \times 400 \times 10^{-6}$$

$$= 1.9 \times 400 \times 10^{-6}$$

Change in resistance $\Delta R = \mathbf{0.152}\ \Omega$

Change in output voltage,

$$\Delta V_o = V_i\left(\frac{\Delta R_1}{R_1 + R_2}\right) = 4\left(\frac{0.152}{200 + 200}\right)$$

Change in output voltage $\Delta V_o = 0.00152\ V = \mathbf{1.52\ mV}$

Figure 3.13 Example calculation determining the change in output voltage of a strain gauge bridge

Figure 3.13 shows an example calculation involving change in the resistance of a strain gauge ΔR and its output voltage ΔV_o.

When strain gauges have been bonded to an object, it is nearly always permanent and they cannot be reused. Their accuracy depends on the skill with which they have been installed. Correct positioning and proper bonding of the gauge is important. Also there are many different types of gauge suited to various strain ranges and environments, and it is important to choose the correct type of gauge. For example, temperature significantly affects the accuracy of some gauges, so these would not be suitable for use in a situation where temperature often rises and falls, or would need electronic circuitry to lessen the effect.

Bonded resistance strain gauges are in widespread use in industry. They are often used in groups, for example to measure strain occurring simultaneously in different directions, or to offset the effects of an object which bends as well as stretches. They are used in civil engineering to monitor strain in road and rail bridges, walls of tall buildings, or embedded in roads to record use or wear. In mechanical engineering they are used extensively to measure strain in metal test specimens, or for example in prototype turbines, cylinders, aircraft, or other machinery.

There are other types of strain gauge available apart from bonded resistance strain gauges. For example, some types of strain gauge are designed to be unbonded, or to specifically measure vibration, or even to detect cracks in structures. Apart from measuring strain and displacement, strain gauges are also widely used to measure force, pressure, or acceleration.

Variable area capacitors

This method of measuring displacement uses the electrical property of capacitance.

Key fact

Capacitance is the property of a system that enables it to store electrical charge. A capacitor is an electrical component having capacitance. Capacitors are formed by conductors separated by a dielectric.

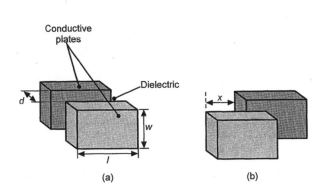

Figure 3.14 Change of plate overlap area

Figure 3.14 represents the conductors of a capacitor, in the form of parallel plates, with a dielectric material between them. A dielectric material is a substance such as mica, glass or kerosene that can sustain an electric field.

The capacitance is measured in farads (F) and is given by the expression:

$$C = \frac{A\varepsilon_0\varepsilon_r}{d}$$

where:

- A is the area of overlap between two capacitor plates (m^2)
- ε_0 is the permittivity of free space ($8.854 \times 10^{-12}\ F.m^{-1}$)
- ε_r is the relative permittivity of the dielectric between the capacitor plates (no units)
- d is the distance between the plates (m)

Relative permittivity is the ratio of electric flux density produced in a material to the value in free space produced by the same electric field strength.

Key fact

Permittivity is the property of a material which describes the electric flux density produced when the material is excited by an e.m.f.

Capacitive displacement transducers produce a change in capacitance proportional to a change in displacement. There are three basic ways in which they can do this. The capacitance C can be changed by varying either the area of overlap A, the relative permittivity of the dielectric ε_r, or the distance d between the plates. We shall consider the type where the area of overlap between the two conductive plates is altered, known as the variable area type.

In Figure 3.14(a) the area of overlap A is $w \times l$. In Figure 3.14(b), the area of overlap A has been reduced by $w \times x$ metres^2 and is now $A - (w \times x)$ metres^2. If the capacitance in Figure 3.14(a) is

$$C = \frac{A\varepsilon_0\varepsilon_r}{d}$$

then the capacitance in Figure 3.14(b) will be

$$C = \frac{(A - wx)\varepsilon_0\varepsilon_r}{d}$$

The example calculation shown in Figure 3.15 demonstrates how changing the area of overlap A changes the capacitance C. This change in capacitance is proportional to the change in area of overlap of the plates. Therefore, by using an appropriate capacitor design, mounting arrangement and measurement system, we can use capacitance change to measure displacement.

An example of a variable area type capacitive displacement transducer is shown in Figure 3.16. It consists of a static cylinder into which a sliding cylinder is inserted to form the capacitive plates. The inside of the static cylinder is coated with dielectric material. The sliding cylinder attaches to the object whose displacement is to be measured via a coupling or plunger. As the sliding cylinder moves in or out of the fixed cylinder, the area between the plates changes, so the capacitive output changes in proportion to the displacement.

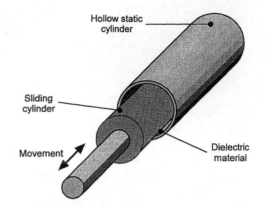

Figure 3.16 Variable area capacitance displacement transducer

---

Problem

Consider a capacitor consisting of two parallel conductive plates in parallel. Each conductive plate has a width w of 0.1 m and length of 0.5 m. The distance d between the two plates is 0.1 m. The relative permeability of the dielectric ε_r is 1. Given that the permittivity of free space ε_0 is 8.854×10^{-12} F.m^{-1}, determine the capacitance of this device. If the overlap of the plates is reduced by moving one plate horizontally a distance x of 50 mm, determine the new value of capacitance.

Solution

From the question we know that w = 0.1 m, l = 0.5 m, d = 0.1 m, ε_r = 1 and ε_0 = 8.854×10^{-12} F.m^{-1}. The area of each plate A is

$$A = wl = 0.1\,\text{m} \times 0.5\,\text{m} = 0.05\,\text{m}^2.$$

To find the capacitance

$$C = \frac{A\varepsilon_0\varepsilon_r}{d}$$

$$= \frac{0.05 \times 8.854 \times 10^{-12} \times 1}{0.1}$$

$$= 4.427 \times 10^{-12}\,\text{F}$$

$$C = 4.427\,\text{pF}$$

If the length of the overlap is reduced by moving one plate horizontally a distance x = 50 mm.

$$\text{Area of overlap} = (A - wx)$$

$$= 0.05 - (0.1 \times 0.05)$$

$$= 0.045\,\text{m}$$

Thus the new value of capacitance C_x will be;

$$C = \frac{A\varepsilon_0\varepsilon_r}{d}$$

$$= = \frac{0.045 \times 8.854 \times 10^{-12} \times 1}{0.1}$$

$$= 3.984 \times 10^{-12}\,\text{F}$$

$$C = 3.984\,\text{pF}$$

Figure 3.15 Effect of changing area of overlap A of a capacitor

The output signal from capacitors needs a significant amount of conditioning. They also need circuitry to compensate for temperature changes which affect capacitance and produce errors. The conditioned signal is then in the form of a voltage proportional to displacement, which is connected to a suitable calibrated voltmeter.

Capacitive displacement transducers are generally only suitable for measuring small displacements. Specifications of this type of displacement sensor are available for use in high humidity, high temperature, or nuclear radiated zones. They are very sensitive, have infinite resolution, but they can be expensive and need significant local signal conditioning. For these reasons they tend to be used for specialist applications. Examples of these are surface profile sensing, wear measurement, or crack growth.

Angular displacement

Angular (or rotary) displacement occurs in many devices and machines. Measuring angular displacement is often necessary for assessing machine performance, and is essential in many positioning and control systems. Other parameters can often be deduced from angular displacement. For example, the angular velocity (angular displacement with respect to time) of the wheel on a vehicle is directly proportional to the vehicle's linear velocity. Mounting a suitable sensor on the wheel or final drive of a vehicle measures its rotational velocity, and this information can be displayed directly to the driver via a speedometer, calibrated in terms of linear velocity.

Rotary potentiometer

Rotary or angular potentiometers measure angular displacement. Figure 3.17 shows a typical rotary potentiometer design.

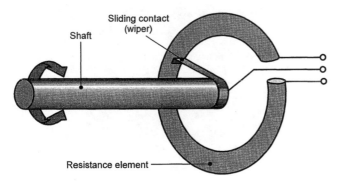

Figure 3.17 Rotary potentiometer

Rotary potentiometers work on the same principle as linear potentiometers, which we met earlier in this chapter. Similar materials and techniques are used, the main difference being that the resistance element is in the form of an arc, which the sliding contact (wiper) follows in an angular motion. The wiper is rotated by the input shaft. The output voltage is proportional to angular shaft displacement, and will usually be displayed on a voltmeter calibrated in units of angular displacement.

The rotary potentiometer shown in Figure 3.17 has a resistive element occupying almost one complete revolution of the shaft. These are known as single turn potentiometers, and are limited because they cannot measure angular displacements of more than one revolution.

Figure 3.18 shows a helical multi-turn potentiometer. The helical potentiometer has a resistive element in the form of a helix

and a mechanical arrangement to allow the wiper to follow the helix as the input shaft is rotated. By this means angular displacements of up to 30 turns can be measured.

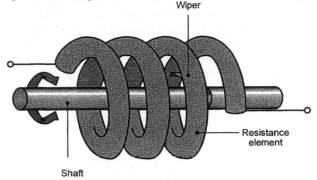

Figure 3.18 Helical potentiometer

Figure 3.19 shows another design of multi-turn potentiometer. This uses a wiper attached to a screw-threaded shaft. The wiper makes contact with a block shaped resistance element. The wiper moves up and down the resistance element as the screw turns.

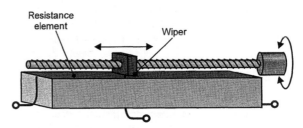

Figure 3.19 Multi-turn potentiometer

The characteristics of rotary potentiometers are generally the same as linear potentiometers. For example, they suffer from wear of the sliding contact and resistance elements, and may create electrical noise when being adjusted. The accuracy of the output will be adversely affected if the former is not a perfect arc, or the wiper is not positioned exactly in the centre of the arc, or if the windings are not uniform. However they are relatively inexpensive and in widespread use.

The ranges of rotary potentiometers are from less than 0.17 radians (10°) to over 61 radians (3500°). They have a linearity from 0.01% to 1.5%. They are used in a wide range of applications, from electronic wheelchairs to aircraft.

Optical shaft encoders

Optical shaft encoders produce information about angular displacement in digital form. This is useful because a digital output is compatible with computers and other digital electronic systems.

> **Key fact**
>
> An optical encoder is a transducer in which linear or angular displacement varies the transmission of light from a source to a detector.

There are two main types of encoder device, the incremental encoder and the absolute encoder. The incremental encoder produces an output signal showing that some displacement of a shaft has taken place. Further output signals are counted and from these the angular displacement of a shaft can be measured, relative to some arbitrary datum. The absolute encoder produces an output signal which shows the total displacement of a shaft from a null position.

Figure 3.20 shows a typical incremental shaft encoder. The incremental shaft encoder consists of a disc rigidly attached to the shaft whose displacement is to be measured. The disc has a number of equally spaced slots, or windows, through which a beam of light can pass. The rest of the disc is opaque. A light source, consisting of two light emitting diodes (LEDs), is aligned with the disc. If the light from the LEDs is uninterrupted, it is detected by the light detectors.

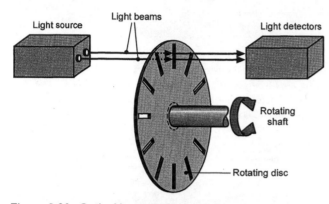

Figure 3.20 Optical incremental shaft encoder

Figure 3.21 shows a typical incremental shaft encoder disc. As the shaft rotates, the light shines through the equally spaced windows in the disc, and is blocked by the opaque sections of the disc. Hence a pulsed light output from the light detectors is produced. The LEDs and detectors are arranged so that, as the disc rotates, the phase difference between the pulse trains from the detectors shows the direction of rotation. The number of pulses detected is proportional to the angle through which the shaft and disc travel. The angular displacement of the shaft can be determined relative to an arbitrarily selected starting point.

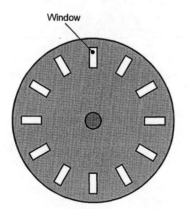

Figure 3.21 Incremental shaft encoder disc design

The resolution of the disc depends on how many windows it contains. The more windows the disc has, the higher the resolution. Resolution is determined by dividing 2π radians or 360° by the number of windows in the disc. The number of windows on the rotating disc can vary from 60 to well over 1000 with multi-tracks, allowing very good resolution to be achieved.

Typical resolutions of optical incremental shaft encoders are 0.0034 radians (0.2°) to 0.102 radians (6°).

Figure 3.22 shows a typical optical absolute shaft encoder. It differs from the incremental encoder in that the output signal it produces is in binary or coded form. This provides an absolute displacement of the shaft.

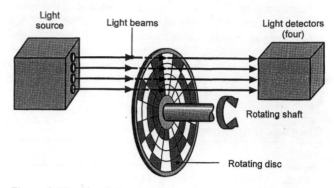

Figure 3.22 Absolute encoder

A rotating disc, with a number of concentric tracks, is attached to the shaft. A light source consisting of LEDs is aligned with the tracks of the disc. A light detector is similarly aligned with the disc and beams to detect the light which passes through the disc.

A 'closed' window, which is opaque and so prevents light from the LEDs reaching the detector, represents a binary '0'. An 'open' window, which allows light from the LEDs through to the detector, indicates a binary '1'. The combinations of open and closed windows follow a binary sequence from 0 to $(2^n)-1$, where n is the number of tracks.

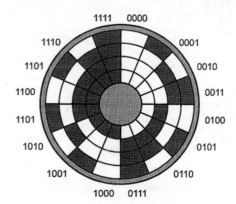

Figure 3.23 Binary absolute encoder disc

The binary absolute shaft encoder disc shown in Figure 3.23 has four tracks and consequently there are four bits in each binary number. The number of positions which can be detected is 16 (which is 2^4), which means the binary sequence runs from 0 to 15 (which is 0 to $2^4 - 1$). The resolution, determined by dividing 2π radians or 360° by the number of windows in the disc (in this case 16), is 0.393 radians or 22.5°. If we use a rotating disc with eight tracks, giving eight bits in each binary number, the number of positions that can be detected is $2^8 = 256$. The resolution is then 0.024 radians or 1.41°.

In practice there are problems with this type of binary absolute encoder. The exact alignment of the window edges in each track is difficult to achieve, and consequently errors are sometimes introduced. These errors occur at the boundaries between windows and, in some cases, it is possible to make a 180° error in determining the shaft angular displacement.

A major disadvantage of the binary absolute encoder is that on many occasions more than one window will change condition for one increment. This is because of the nature of the binary number system. Examples of this are; 0011 to 0100 (3 to 4 in base 10), from 0111 to 1000 (7 to 8 in base 10), and so on. The most significant change is from 1111 to 0000 (15 to 0 in base 10). Hence if the absolute encoder system misreads one window, it can lead to serious errors in position measurements. To overcome this the Gray Code (named after Frank Gray of Bell Laboratories) was developed. This produces a sequence where only one 'bit' or 'window' changes condition between consecutive positions.

Key fact

Gray code is a digital code in which only one bit at a time changes as the value incrementally increases or decreases.

Angular optical encoders have applications in numerically controlled machines, such as computer controlled lathes or milling machines. They are also used in robotics and positioning systems. A common application of relative optical encoders is in computer the computer mouse.

Tachometric generator

A tachometer refers to any device used for measuring the rotation of a shaft (from the Greek word takhos, meaning speed). A generator is a device which converts mechanical energy into electrical energy. A tachometric generator is a machine which, when driven by a rotating mechanical force, produces an electrical output proportional to the speed of rotation.

Tachometric generators connect to the rotating shaft whose displacement is to be measured. There are several methods of doing this, for example by direct coupling, or by means of belts or gears (although any rotational amplification must be taken into consideration). They produce an output which primarily relates to speed and not displacement. However, speed is the rate of change of displacement, and a measure of displacement can be obtained by integrating the tachometer output over time. We shall see how to do this in Chapter 9.

Tachometric generators are generally referred to as a.c. or d.c. tachogenerators, or a.c. or d.c. tachometers.

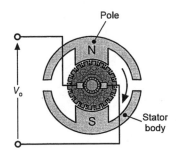

(a)

Figure 3.24 D.C. tachometric generator

Figure 3.24 shows a d.c. tachometric generator. It is essentially a small d.c. generator, which produces a moderate d.c. output voltage. It differs from a d.c. generator in that certain parts of the design are optimised to give greater accuracy as a speed measuring instrument rather than the generation of electricity. Because generators are made largely of conductive and ferrous metals, they can be quite heavy. Tachometers do not always need to be as robust as generators, and they may incorporate other, lighter materials, such as fibreglass, to reduce their total mass. It is important to make these devices as light as possible so the mass of the tachometer does not affect the speed of the system being measured.

The output signal from d.c. tachometers usually requires electronic circuitry to remove electrical noise. The output signal can then be displayed on a voltmeter calibrated in terms of speed or displacement. A characteristic of d.c. tachometers is that the polarity of the output voltage shows the direction of rotation of the shaft. The range of measurement is between about 0–600 radians per second (0–6000 revolutions per minute). They need regular maintenance because certain parts, particularly the brushes, are prone to wear, and also the strength of permanent magnets tends to weaken over time.

As its name suggests the a.c. tachogenerator is an electrical generator that produces an a.c. output.

Figure 3.25(a) shows a typical a.c. tachogenerator arrangement with a permanent magnet (rotor) rotating within a stationary coil (stator winding). In normal operation the rotor would be connected to and driven by the shaft to whose angular speed is to be measured, using the same techniques described for the d.c. tachogenerator. This may be represented schematically as shown in Figure 3.25(b).

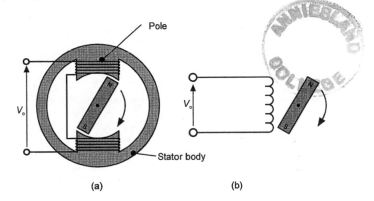

(a)   (b)

Figure 3.25 A.C. tachometric generator

The output, V_o, is an alternating signal, the amplitude and frequency of which are both proportional to the speed of rotation. Using suitable signal processing circuits, either amplitude or frequency may be used to give an indication of speed.

Compared with the d.c. tachometer, the a.c. tachometer has the disadvantage of requiring more involved signal conditioning. Also, the direction of rotation cannot be obtained from the output signal. However, a.c. tachometers are simpler, less expensive, and more reliable. They require less maintenance and, when the frequency is used to measure speed, they give long term accuracy even if the strength of the permanent magnet weakens. Also, the a.c. tachogenerator does not have the high noise content in its output signal which is normally associated with the d.c. tachogenerator.

Both a.c. and d.c. tachometric generators are widely used in automated production systems, machine tools, and for monitoring large electricity generators.

Proximity

Proximity measurement refers to indicating whether an object is present within a defined area close to the sensor. From this, angular or linear displacements, and speed and velocity can be determined. For example, proximity sensors are often used to measure rotation by indicating when a point on a shaft is present. Assuming the shaft rotates in one direction, the number of times the point on the shaft is sensed passing the sensor will show the number of rotations of the shaft, and so its displacement. Hence the proximity sensor is used to measure angular displacement accurately to within one revolution. In a similar manner it can measure velocity and acceleration, by measuring events with respect to time.

Proximity detectors are used widely in manufacturing processes. They often monitor the position, presence, or non-presence of components during assembly. They are also used to count products, for example items on a conveyer belt to be packed in a set size of box.

Microswitches

As their name suggests, microswitches are small switches. They operate with very little movement of an operating plunger, so are sensitive and particularly suitable for use as direct contact proximity sensors.

| Plunger | Lever | Roller |

Figure 3.26 Various microswitch operating attachments

Microswitches are available with a variety of operating attachments, some of which are shown in Figure 3.26. The contacts of the microswitch may be normally open, normally closed, or changeover. This opening or closing of contacts is used to make or break a circuit.

Microswitches are in widespread use in many applications, and are probably the most commonly used proximity sensor. They can be made robust and sealed to withstand different types of environment, and they are relatively simple and inexpensive. They are contact devices however, and require a force, albeit small, to operate them.

Typical applications of the microswitch include safety guards, which switch off machinery or sound an alarm if a cover is opened. The operating attachment can be in the form of a sprung plunger with a roller which follows the profile of a cam. In this way, microswitches can be used in conjunction with a cam or series of cams to control the sequencing in machinery. They are also used in quality control applications, for example to reject or accept a component based on its ability, because of its size or weight, to operate a microswitch or not.

Figure 3.27 shows a technical specification for two types of industrial metal-housed microswitches. Note the significant lowering of voltage and current for d.c. compared to a.c. This is to allow for the tendency for d.c. to arc on contact opening, and is typical of all switches.

Industrial metal housed microswitches, unsealed

Contact rating
480 V a.c. 15 A, 125 V d.c. 0.5 A, 15 V d.c. 15 A.

Mechanical life
$>10^5$ operations

Temperature range
0–70 °C

Operating limits	Plunger type	Roller type
Pre-travel max.	0.5 mm	0.8 mm
Differential max.	0.08 mm	0.08 mm
Over travel min.	4.6 mm	3.6 mm
Operating force max.	4.2 N	4.2 N
Release force min.	1.7 N	1.7 N

Figure 3.27 Industrial microswitch specifications

Variable reluctance proximity sensors

Variable reluctance proximity sensors (often referred to as magnetic pick-ups) are small magnetic devices often used to detect angular displacement. The sensor consists of a small electromagnetic coil held in a protective casing, mounted in a fixed position close to the shaft. It can detect the presence of a ferrous metal, for example a ferrous gear tooth.

Figure 3.28 shows a variable reluctance proximity sensor detecting the immediate presence of a gear tooth. When the tooth passes close by the pick-up, an output voltage is produced caused by variations in its magnetic field. The output is a pulse, and may be displayed on a voltage or current meter. The angular rotation of a shaft can be determined by embedding a suitable ferrous metal in a shaft, and counting how often an output voltage is produced.

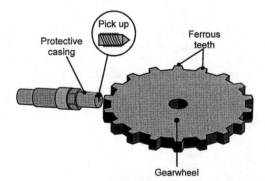

Figure 3.28 Variable reluctance proximity sensor

Typical variable reluctance proximity sensors can detect ferromagnetic materials up to 2.5 mm away. They are widely used in applications such as crank-angle sensing and ignition timing in engines, in disk drives in computer systems, and speed sensing in motors. They can be made very small and so deployed where other sensors may not fit. Many devices of this type are passive devices, requiring no external power. They are often sealed in protective cases and can be highly resistant to environmental influences, such as extremes of temperature and pressure, or chemical attack. They are low cost, but have to be very close to a suitable ferrous metal to produce an adequate output voltage. They also suffer from unwanted signals, or noise, and generally only work at medium to high speeds.

Hall-effect proximity sensors

Hall-effect sensors are magnetically operated proximity detectors. They can detect very weak magnetic fields or small changes in magnetic flux density.

Key fact

The Hall effect is the generation of a transverse voltage in a conductor or semiconductor carrying current in a magnetic field.

Figure 3.29 shows the basic principle of the Hall effect. When a current carrying conductor or semiconductor is acted upon by a magnetic field positioned at right angles to it, the electrons passing through the conductor are acted upon by a force which concentrates them more to one side of the conductor than the other. This variation in current distribution creates an e.m.f. across the conductor. This e.m.f. is proportional to the strength of the magnetic field, and so can be used as the principle of a proximity sensing device.

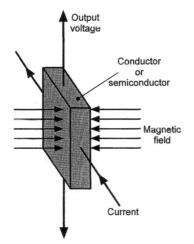

Figure 3.29 The Hall effect

Hall-effect devices tend to use semiconductors, because the effect is more pronounced. Hall-effect proximity sensors consist of a small integrated circuit encased in a probe, which can accurately detect the movement of ferrous metal targets.

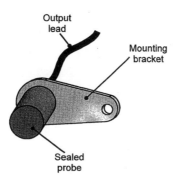

Figure 3.30 Hall-effect proximity sensor

A typical Hall-effect proximity sensor design is shown in Figure 3.30. They are used in much the same way as variable reluctance proximity sensors, for example for monitoring the proximity of ferrous gear teeth. They tend to be more expensive than variable reluctance proximity sensors, but have much better signal to noise ratios, and are suitable for low speed operation.

There are many other devices based on the Hall effect, for example, Hall-effect vane switches, Hall-effect current sensors, and Hall-effect magnetic field strength sensors.

Optical sensor

Optical sensors consist of a light source and a detector. Figure 3.31 shows the principle of a reflective optical beam sensor, where the light source and detector are mounted side by side. Another optical method is known as transmissive or through-scan, where the light source and receiver are separate and aimed at each other.

Light sources are often light emitting diodes (LEDs) and detectors silicon phototransistors (a semiconductor device, whose properties change when light falls on it). Visible red or infrared light is usually used. Visible red light often makes installation and maintenance easier, but infrared suffers less from interference from other light sources.

The source emits visible red or infrared light which is reflected by any object approaching the sensor. The reflected light is detected by the phototransistor.

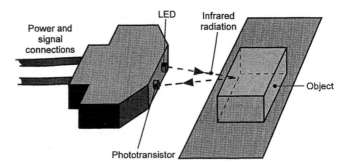

Figure 3.31 Reflective optical beam sensor

In through-scan or transmissive beam sensors the light beam is interrupted, so no light reaches the detector indicating the presence of an object. With the reflective method, the presence or strength of the reflected beam indicates the proximity of an object. To reduce errors, the object whose presence is to be detected may have a special reflective label or coating. The light may also be transmitted in pulses and filtered to prevent false readings.

The minimum proximity distance at which the sensor will operate depends on the power of the LED, the sensitivity of the phototransistor and the nature of the reflecting object. For the reflective method it ranges from less than one mm to over 7 m. Fibre optics can be used to allow this technique to be applied in places inaccessible to other methods of proximity measurement.

The light detector unit is often self contained, producing an appropriate voltage output showing whether the beam has been broken, or on more advanced models, giving an indication of distance. The output voltage will usually require amplification, but it allows for remote reading.

Example applications of optical sensing devices are in alarm systems, and for quality control, particularly in mass production situations.

Reed switch sensors

Figure 3.32 shows a reed switch sensor. Reed switches consist of two small ferromagnetic reeds hermetically sealed in a glass tube. The reeds are thin, and flexible, and because they are ferromagnetic they become magnetised in the presence of a magnetic field. The reeds may have normally open or changeover contacts, depending on the application.

If the reed switch is the normally open type, when a magnet passes close to the reed switch, the ends of the reeds are attracted together and complete a circuit. When the magnet is removed the switch opens again. The changeover type has one flexible reed between two contacts. When a magnet passes close to the reed switch, the reed moves from the normally closed contact to the normally open contact, reverting to normal when the magnet is removed.

Key fact

Hermetically sealed means something that is sealed so as to be airtight.

The angular rotation of a shaft can be determined using a reed switch by embedding or attaching a permanent magnet to a shaft. The reed switch is fixed near the shaft, so that when the shaft rotates the magnet passes close by. As the magnet passes, the reed switch makes or breaks a circuit and produces a pulsed output. These pulses can be used to determine angular displacement, velocity, or acceleration using appropriate signal conditioning.

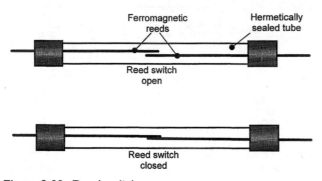

Figure 3.32 Reed switch sensor

The distance the magnet needs to be from the reed switch before it operates depends on the strength of the magnet, the material the reeds are made of, and the distance between the reeds when the switch is open. The operating speed of the reeds can be less than 1 millisecond. Because the tube is hermetically sealed, impurities such as dust do not affect the reeds and so magnetic reed switches have a long working life and need little maintenance compared to other switching devices. However, they are fragile and may require protection against knocks or vibration.

Reed switches are relatively inexpensive, and a typical application is in bicycle speed and distance computers. Here, a small magnet is attached to the one of the spokes of a bicycle wheel, and the reed switch is fixed to the bicycle frame or forks. As the magnet passes the reed switch, a pulsed output is produced. Because the circumference of the wheel is known, a small computer with a liquid crystal display shows the output to the rider in terms of the linear speed of the bicycle or the distance it has travelled.

Other applications of reed switches are in interlock safety guards, which automatically break a circuit to shut a machine off when a guard is removed, and in switches on doors and windows for burglar alarms.

Acceleration

Acceleration is the rate of change of velocity. Accelerometers are used for two specific types of measurement. Firstly, they measure shock and vibration. Shock is effectively a large acceleration over a short period of time, and vibration is a short acceleration regularly repeated. Secondly, accelerometers are used for measurement of the acceleration of bodies, such as aircraft and ships, providing information such as position, speed, and distance travelled. Acceleration is measured in metres per second per second, but is sometimes expressed in terms of the acceleration due to gravity, g. The value of g varies slightly depending on where it is measured, but one g is generally accepted as being 9.80665 m.s^{-2}, usually simplified to 9.81 m.s^{-2}.

Because the acceleration of an object is equal to the force acting on it divided by its mass, accelerometers may measure force and, because the mass is known, determine acceleration from it. They may also determine acceleration from displacement, as with the seismic mass accelerometer.

Seismic mass accelerometers

Previously we have seen how some types of transducer measure one type of parameter to determine another, related parameter. A seismic mass accelerometer measures the linear displacement of a mass from a datum, to determine acceleration.

When a mass accelerates, it experiences a force proportional to its acceleration. Also, when a force acts upon a mass, the mass will be displaced by a distance proportional to the force acting on it and any other forces opposing it. For example, if we are in an aeroplane when it is accelerating along a runway to take off, we feel a force pushing us back into our seats. If the seats were not fixed to the aeroplane, the same force would move the seat backwards. The amount we are pushed into the seat is proportional to the acceleration of the plane and the forces opposing our displacement. Seismic mass accelerometers use this principle to measure acceleration in terms of displacement.

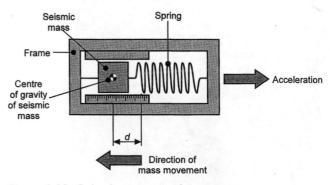

Figure 3.33 Seismic mass accelerometer

Figure 3.33 shows a seismic mass accelerometer. In this sense, seismic mass refers to a mass used as a reference. This mass, m, is attached to a spring of a known stiffness, K. The spring is attached to a frame or case. When the frame is accelerated, the mass is displaced a distance d relative to the frame in the opposite direction. By measuring d we can determine the acceleration. The acceleration, a, is given by:

$$a = \frac{dK}{m}$$

where:

- a is the acceleration in m.s^{-2}
- d is the displacement of the mass relative to the frame in m
- K is the stiffness of the spring in N.m^{-1}
- m is the size of the mass in kg

The higher the acceleration, the greater the displacement of the mass. When the frame stops accelerating, the mass returns to its original position. When the frame decelerates, the mass compresses the spring.

In practice the displacement of the mass is measured with a device such as an LVDT or linear potentiometer, and the acceleration calculated by a computer.

Seismic mass accelerometers are used to measure shock and very low frequency vibrations, such as those related to earth tremors, or measuring the effect of underground detonations.

Piezoelectric accelerometer

The piezoelectric accelerometer is one of the most popular types of accelerometer. Figure 3.34 shows a piezoelectric accelerometer. It consists of a mass attached to a piezoelectric crystal, mounted inside a casing.

The mass is held by a screw against the crystal. When the mass accelerates, it applies a force on the crystal. A charge then occurs in the crystal, and polarises on the metallised faces. The metallised faces connect to an electronic circuit.

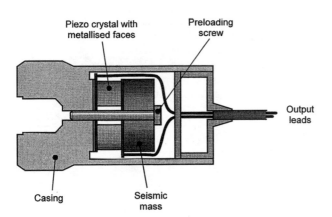

Figure 3.34 Piezoelectric accelerometer

Figure 3.35 shows a typical specification for a piezoelectric accelerometer.

The circuit, known as a charge amplifier, produces a useful output voltage from the charge in the crystal. This output voltage is proportional to the acceleration of the transducer, and can be displayed on a calibrated voltmeter.

Key fact

The piezoelectric effect, discovered by Pierre and Jacques Curie in 1880, causes electric charges of opposite polarity to appear on the faces of certain types of crystal when subjected to a mechanical strain. This charge is proportional to the strain.

Crystals such as quartz and sodium potassium tartrate (rochelle salt) are used because they have high mechanical strength. They are also relatively inexpensive.

Industrial process control piezoelectric accelerometer with built in signal conditioning
Output
4 to 20 mA
Frequency response
2 Hz to 1 kHz ±10%
Measurement range
50 g peak
Sensitivity
0–10 mm.sec^{-1}
Operating temperature range
–25°C to 80°C
Temperature sensitivity
0.08 $\text{\%.°C}^{-1}$
Electrical noise
0.3 mV max.
Supply voltage
10–32 volts
Weight
150 gms

Figure 3.35 Typical piezoelectric accelerometer specification

Piezoelectric accelerometers have good high frequency response, but poor low frequency response. They produce relatively high output voltages but this is often accompanied by significant electrical noise. The transducers are small and light in weight and can withstand very high accelerations, of up to $250000\ \text{m.s}^{-2}$. Typical applications are where high frequency, high accelerations occur, such as in impact testing.

Summary

In this chapter we have looked at motion, and devices and techniques commonly used to measure it. Measuring motion can often quantify other parameters in a system, so devices that measure motion, particularly displacement, form the basis of many other types of transducer.

All the devices we have looked at are available in varying designs and sizes. Of all the measurement parameters, there are probably more devices, methods and techniques available to measure displacement, velocity, or acceleration, than any other physical quantity.

There are many other devices, techniques, and adaptations of the ones we have discussed, available to measure the various forms of motion. However, the most commonly used devices, which often encompass the basic principles or concepts of other methods, have been included. This chapter will have provided you with an insight into the subject of motion measurement, and a good basis to understand other types of motion sensor you may meet.

Questions for further discussion

1. When using optical encoder methods to measure rotary displacement, describe the type of problems you may be experiencing to consider changing from a binary absolute encoder to an encoder using Gray code.
2. What specific properties need to be possessed by the adhesive used to bond a strain gauge onto an object?
3. Describe a practical situation where using Hall-effect proximity sensors would be more suitable than variable reluctance proximity sensors. Give reasons for your answer.
4. In a manufacturing plant, think of a practical situation where using a linear potentiometer to measure displacement would be more suitable than an LVDT. Give reasons for your answer.
5. Consider the magnetic reed switch used as part of a speed and displacement sensor system on a bicycle. What problems or disadvantages could you foresee if the reed switch were to be replaced by a variable reluctance proximity sensor?
6. Consider the seismic mass accelerometer shown in Figure 3.23. This device can be modified to use strain gauges to measure the displacement of the seismic mass. Discuss how this might be achieved, and any advantages of using strain gauges for this application.

4 Level, Height, Weight and Volume Measurement

In this chapter we will consider different types of sensors commonly used to measure the quantity of a substance.

Quantity may be expressed in terms of level, height, volume, weight or force, which are different physical parameters but all related.

To begin to understand the relationship between level, height, volume, weight and force, consider the container shown in Figure 4.1. This is a simple, uniform shape container made of a clear material such as glass or plastic (like the measuring cylinders found in science laboratories), designed for holding up to one litre of liquid.

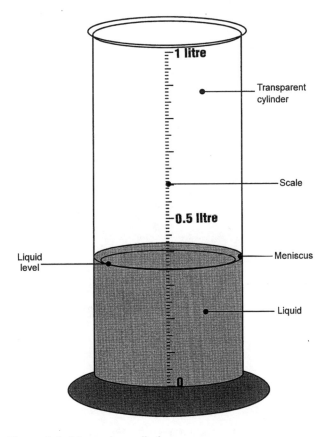

Figure 4.1 Measuring cylinder

When the container is empty, the level of liquid is zero. This is indicated on the side of the container and easily seen. Similarly, when the vessel contains one litre of liquid (full up), the level corresponds to the one litre mark on the side of the vessel and is easily observed. (Note the level should always be read from the bottom of the meniscus).

Because the vessel is of uniform cross-sectional area, the scale can be marked at equal distances up the side of the container corresponding to convenient quantities of liquid. This type of scale is linear – that is, equally spaced, and each mark is called a calibration mark, because it corresponds to a defined quantity (in this case one litre or fractions of one litre).

A linear scale is useful when estimating a level measurement which lies between calibration marks. Alternatively, the container could be calibrated so each mark corresponds to the depth or height of liquid, in millimetres for instance. The marks could also be calibrated in units of volume, such as mm^3. The choice of the units of calibration depends on the information needed, determined by the application.

In practice, where the dimensions of the container are known the calibration mark positions are usually calculated. Sometimes it is convenient to calibrate the container by adding known amounts of liquid and marking the level at each increment. The care taken in positioning the calibration marks contributes to how accurate the measurement system is. The accuracy in this case also depends on the skill of the reader.

The measuring cylinder is an example of a method of measuring the amount of a liquid in a vessel, which can be calibrated to several parameters. The same principle can be applied to many designs of container or tank, for many purposes and of varying designs. If, however, the cross-sectional area of the container varies with the level of liquid, the scale will be non-linear.

Something else happens to the vessel as the quantity of liquid in it increases – its mass increases and it gets heavier because of gravitational force.

If the same vessel is filled with one litre of water then it will increase in mass by 1 kilogram (kg) and in weight by 9.81 newtons (N). If the scale is calibrated in mass, then using a liquid with a different density will require the scale to be recalibrated. If, for instance the same container is filled with mercury instead of water the mass would increase to 13.6 kg because mercury is 13.6 times more dense than water.

Also, the density of any liquid changes with temperature, usually becoming lower as temperature increases and the liquid expands. In most measuring vessels the calibrated scale includes a temperature at which the measurement may be taken without adjusting the value observed.

The physical parameters of level and height can be related to those of mass and so lead to force or weight. In this context all of these physical parameters are directly affected by the amount of material, solid, liquid, or a gas, in a container. This chapter discusses the commonly used sensors for measuring these parameters.

Level measurement

As previously discussed, the physical parameters of level, height, volume, weight or force are all related, and by measuring one of these parameters we can often determine another. The following devices all measure level, but they may be calibrated in terms of one of the other parameters.

Sight glass

The sight glass, shown in Figure 4.2, is a simple and inexpensive method of measuring the level of a liquid in a container. It is similar in principle to the measuring cylinder, but allows opaque materials to be used for the tank construction and hence stronger or less expensive designs.

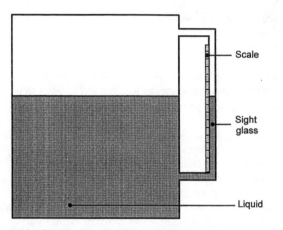

Figure 4.2 Sight glass

The precision and accuracy of the result depends on the skill of the reader, and how accurate and precise the scale gradations are. The type of liquid in the tank and the diameter of the sight glass will also affect the accuracy and precision of this method. Hence sight glasses may be used in applications where high precision and accuracy are not vital, such as in oil storage tanks, or domestic kettles

Dipstick

The dipstick, shown in Figure 4.3, is a simple and inexpensive (if not precise) method of determining liquid level. The dipstick consists of a thin rod with a graduated scale, which in normal use is inserted vertically into a tank so that its lower end is submerged below the level of the liquid. On withdrawal of the rod from the tank a film of the liquid adheres to the rod, allowing the level to be read from the graduated scale.

This measuring instrument is essentially the same as the calibration scale on the side of the measuring vessel. Instead of being permanently attached to the side of the container, the scale is on a removable probe, rod, or stick.

For some applications, such as dipsticks used in petrol tankers, the scale is calibrated in volume to show the amount delivered. In the case of the dipstick used in a car engine, the scale usually has only two marks – MAX (for maximum because too much oil is harmful to the engine) and MIN (because lack of oil will cause damage). As long as the oil level is between the two limits then there is an acceptable amount in the engine.

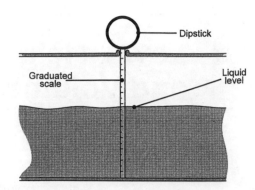

Figure 4.3 Dipstick

The dipstick is for local measurement only, in that someone has to physically remove the dipstick to make a reading.

Float operated gauges

There are two common forms of float operated gauge – the counterweight and the electrical version.

The counterweight operated float gauge

A typical counterweight operated float gauge is shown in Figure 4.4. Here, the movement of the float follows the changing liquid level and this is relayed to the pointer. The scale can be calibrated in volume or mass.

Accurate readings can be obtained but this depends upon scale length and the extent of scale graduations.

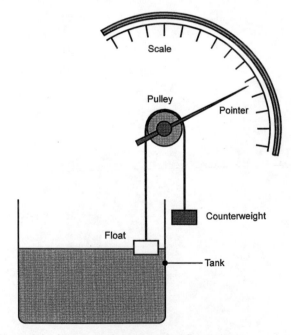

Figure 4.4 Counterweight operated float gauge

Electrically operated float gauge

A typical electrically operated float gauge is shown in Figure 4.5. The float is designed to follow the level of the liquid. When the level changes, the movement of the float produces an angular movement of the slider along the potentiometer. This alters the potential difference, producing a voltage reading that is directly related to the level of the liquid.

The scale of the voltmeter can be calibrated in volume, mass, or height.

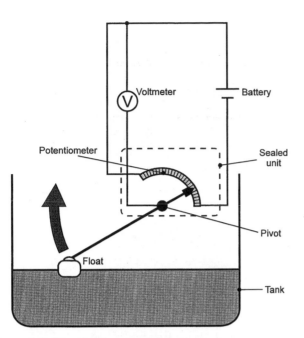

Figure 4.5 Electrically operated float gauge

Because the level signal is electrical, it may be conditioned for remote recording or display, or used as a feedback signal in a control system.

Capacitance probes

Because of their inherently safe operation, capacitance probes (sometimes called capacitance gauges) are regularly used to measure fuel quantity in aeroplanes.

The sensor comprises two cylindrical tubes, one inside the other, which form a capacitor in the fuel tank as shown in Figure 4.6. The gap between the inner and outer plates is normally air, and this is replaced by fuel as the level increases.

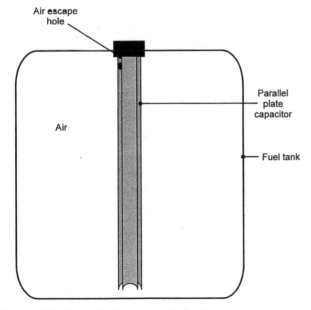

Figure 4.6 Capacitor in an empty fuel tank

If the tank is filled with fuel, as in Figure 4.7, the capacitance will change since fuel has a higher dielectric value than air.

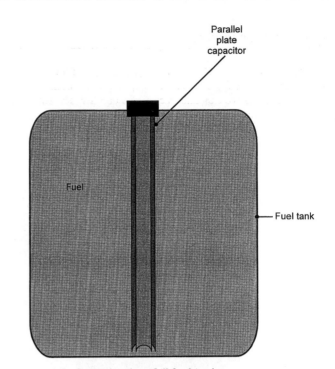

Figure 4.7 Capacitor in a full fuel tank

If the tank is partly filled, as in Figure 4.8, the capacitance will change proportionally with the level of fuel in the tank.

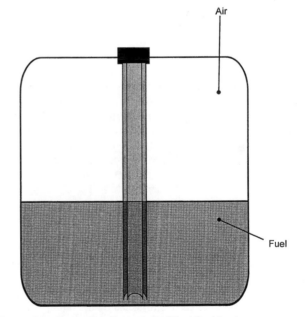

Figure 4.8 Capacitor in a partly filled fuel tank

As the fuel level rises and falls then the added value of capacitance due to fuel, also rises and falls. Any change in capacitance can be used to produce a change in electrical output, which can then be used to operate a fuel level or quantity indicator, or be linked to a control device.

Because of the presence of the fuel, the design of the capacitance probe needs to take into account criteria such as corrosion and leakage.

Conductance probe

Another method of liquid level measurement, similar in principle to capacitance gauges, is the conductance probe.

Instead of measuring change in capacitance, the conductance probe measures the change in resistance of an electrically conductive liquid.

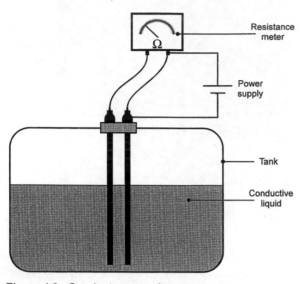

Figure 4.9 Conductance probe

A typical conductance probe, as shown in Figure 4.9, senses the change in resistance between two electrodes as the liquid level changes. The meter measuring the change in resistance can be calibrated in suitable units to act as a level or quantity indicator, or link to a control device.

Corrosion, leakage, and the conductivity of the liquid being measured are important factors to consider when using a conductance probe. Environmental changes such as temperature will affect conductivity of the liquid and the measurement system, as can impurities or changes in type or composition of the liquid.

Because a potential difference is present, in some cases arcing between the tips of the probes may occur if they become uncovered. This is important when considering a level measurement device for flammable liquids.

Ultrasonic level indicator

Figure 4.10 shows a typical ultrasonic level indicator configuration.

Pulses of ultrasonic sound waves are emitted towards the liquid and a small portion of each is reflected by the liquid surface. The rest of the pulse is reflected by the base of the liquid container. Both reflected pulses can be displayed on a cathode ray oscilloscope or other timing based display or recording device.

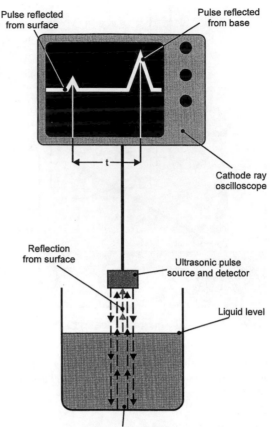

Figure 4.10 Ultrasonic level indicator

The difference in time, t, between the two pulses is directly related to the depth of the liquid being measured and this allows the display to be calibrated with respect to height (or depth or volume).

Although relatively expensive this method is precise and accurate, and can be used over a wide range of depths. Consequently it has applications in many fields from marine depth measurement to medical equipment, and is not just confined to liquid level measurement.

Bubbler level gauge

Figure 4.11 shows a diagram of a bubbler level gauge. The air pressure in the bubbler tube is adjusted via the regulating valve until air bubbles are just beginning to leave the bottom of the bubbler tube.

The gauge meter can be calibrated in units of height or volume, since the pressure at which bubbles will emerge is directly proportional to the height, h, of the liquid in a regular sided container.

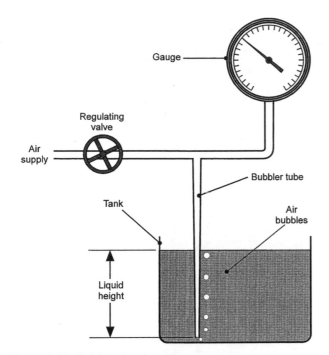

Figure 4.11 Bubbler level gauge

Precise and accurate readings can be obtained by this method, although, like the dipstick, measurement requires human intervention.

Pressure sensors

In certain cases where weight or level measurements are taken and an electrical output signal is required, pressure sensors may be used. Pressure sensors are discussed in detail in Chapter 7. Various types may be used, but a typical example would be piezoresistive.

In a pressure sensor, pressure is used to deflect a diaphragm. The diaphragm deflection generates an electrical signal, via a device such as a resistance strain gauge, giving a signal corresponding to pressure.

Pressure sensors are located in a container in which the height or depth of the liquid is to be measured. Any change in the height or depth of the liquid produces a proportional change in the sensor output. Referring to Figure 4.12, pressure P is:

$$P = \rho g h,$$

then

$$\text{Height}, h = \frac{P}{\rho g}$$

where:

- ρ is the density of the liquid in $kg.m^{-3}$
- g is the acceleration due to gravity, that is, $9.81\ m.s^{-2}$
- h is the height of the liquid, in m

Therefore h is proportional to P since ρ and g are constant for a given liquid in a container of regular cross-section.

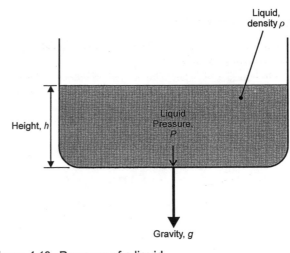

Figure 4.12 Pressure of a liquid

Measurement of weight or force

The weight of a substance is its mass multiplied by the acceleration of gravity. It is often easier or more convenient to measure the weight of a substance and determine other parameters, such as its volume or level, from this. The following devices all measure force, but, similar to level measuring devices, they may be calibrated in terms of other parameters.

Load cells

Load cells are systems that usually use strain gauges to provide measurements of unknown force (sometimes termed load) or mass.

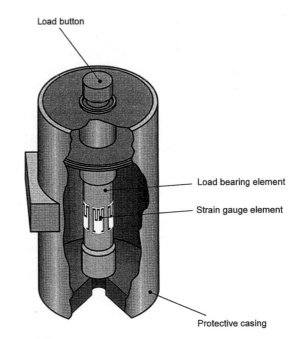

Figure 4.13 Strain gauge load cell

A typical strain gauge load cell is shown in Figure 4.13. It uses four strain gauge elements bonded on to the load bearing element. When the load bearing element is under strain, small changes in resistance occur in the strain gauge elements.

The elements, along with resistors of fixed value, will be incorporated into a Wheatstone bridge arrangement, the precise nature of which depends on the application and the type of strain being measured. A typical Wheatstone bridge arrangement is shown in Figure 4.14, where one or more of the resistances R may be a strain gauge element. The Wheatstone bridge conditions the signal to read the appropriate form of strain. The value of the potential difference at the output of the Wheatstone bridge is related to the magnitude of the applied load.

Errors due to temperature (which affects resistance) can be minimised by electrically matching the strain gauges.

Wheatstone bridges are discussed in more detail in Chapter 9.

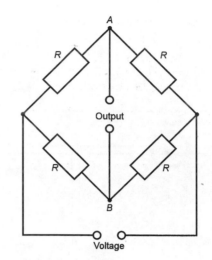

Figure 4.14 Wheatstone bridge

Strain-Gauge Load Cells: compression, tension, and universal.
Compression only precision load cell.
Tension only precision load cell.
Universal (either tension or compression) general purpose load cell.

Electrical data (all models)

Rated output (RO)	3 mV/V
Recommended excitation	10 V a.c. or d.c.
Maximum excitation	20 V a.c. or d.c.
Zero balance	±1% RO
Terminal resistance–input	350 ±3.5 Ω
Terminal resistance–output	350 ±5.0 Ω
Insulation resistance (bridge to ground)	5000 MΩ

Mechanical data

	Capacity (N)			
	4460	8920, 13,380	22,300	44,600
Mass (kg)	2.7	2.7	4.5	4.5
Deflection (mm)	0.15	0.13	0.18	0.13
Natural freq. (Hz)	1800	2600	2200	3400
Length (mm)	115.90	115.90	150.83	150.83
Diameter (mm)	88.90	88.90	88.90	88.90

General data

	C3P1	T3P1	U3G1
Accuracy			
Calibration accuracy (% RO)	0.1	0.1	0.25
Non-linearity (% RO)	0.05	0.05	0.10
Hysteresis (% RO)	0.02	0.02	0.02
Repeatability (% RO)	0.02	0.02	0.02
Temperature			
Temperature range, compensated	From –9 °C to 46 °C		
Temperature range, safe	From –34 °C to 79 °C		
Temperature effect on rated output (% reading/°C)	±0.0014	±0.0014	±0.009
Temperature effect on zero balance (% RO/°C)	±0.0027	±0.0027	±0.0027
Adverse load rating			
Safe overload (% rated capacity)	150	150	150
Ultimate overload (% rated capacity)	300	300	300
Maximum side load (% rated capacity)	30	10	10
Maximum bending moment (% rated capacity, in Nm)	2.8	2.8	2.8
Maximum torque load (% rated capacity, in Nm)	1.1	1.1	1.1

Figure 4.15 Strain gauge load cells specification

Load cells are used, for example, in measuring:

- The weight of material stored in bins.
- The weight of vehicles on weigh-bridges.
- The level or volume of liquids in tanks (if the weight is known then the volume or level can be calculated).

This technique is accurate, robust, relatively inexpensive and can be used over a wide range of loads.

Figure 4.15 shows the specification details of a particular range of strain gauge load cells. Variants are available for use in compression only, tension only, or for response to either direction of load.

The specification details show, for example, that these devices are:

- Accurate (better than 0.25%).
- Linear (better than 90%).
- Have good repeatability (better than 0.25%).
- Resilient to overload (better than 150%).

We shall meet these devices again in one of our application examples in Chapters 10.

Balance and scales

Balance and scales are force measuring systems which use balancing to determine unknown values of mass and force (weight).

Referring to Figure 4.16, the gravitational force acting on the unknown mass W_x is compared with that of the known mass W. The balance is a lever with the known and unknown forces being of equal distance d from the pivot. Balance condition ($W_x = W$) is achieved by adding known masses to the right-hand pan until the beam remains horizontal.

Using a balance and scales with a sliding mass, as shown in Figure 4.17, a coarse balance is achieved by adding known masses to the balancing pan. An accurate result is then obtained by moving the slide weight, W_s, to the optimum position for retaining a horizontal beam (fine tuning).

Because $(W_x \times a) = (W \times b) + (W_s \times c)$ and W_s is constant, the distance c can be calibrated directly. Hence the horizontal beam will usually have a calibrated scale along the path of the sliding mass.

To increase or decrease the range of measurement, the range of known masses available for the balancing pan is changed accordingly.

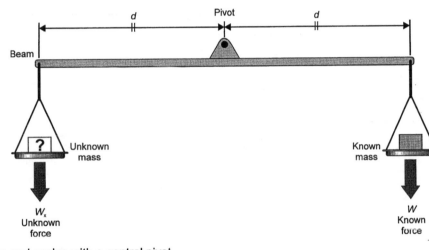

Figure 4.16 Balance and scales with a central pivot

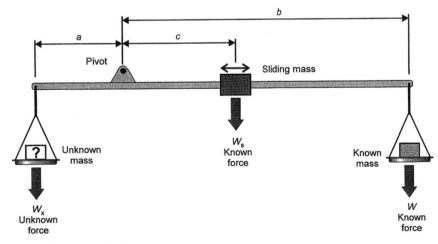

Figure 4.17 Balance and scales with a sliding mass

Spring balance

A typical commercially available spring balance is shown in Figure 4.18. The spring is anchored and stretched by the applied unknown force W_x.

The extension of the spring is directly proportional to the force applied (Hooke's Law). This allows the scale to be calibrated in units of force (Newtons) or in units of mass (kilograms).

> **Key Fact**
>
> Hooke's Law states that for an elastic material the strain is proportional to the applied stress. (The value of the stress at which a material no longer obeys Hooke's Law is known as the Limit of Proportionality).

High precision or accuracy cannot be achieved by the spring balance method, but it does provide a quick and easy estimation of mass or force.

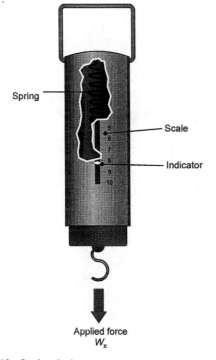

Figure 4.18 Spring balance

The range of mass or force that can be measured, by a particular balance, depends upon:

- The material from which the spring is made.
- The thickness of the spring material.
- The diameter and length of the spring.

By combining the spring balance with a potentiometer (as shown in Figure 4.19) an electrical output signal can be produced which is proportional to the applied load.

The spring balance converts the applied force into displacement and the potentiometer converts this displacement into an electrical output signal. This is a simple and inexpensive method of obtaining remote readings of mass, force, level, volume and pressure. In this case the potentiometer produces an output signal which is proportional to force. Note that it is movement (displacement) which is causing the actual change in the output signal.

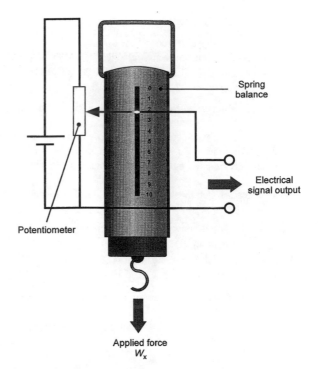

Figure 4.19 Spring balance and potentiometer

This is an example of one sensor type being able to measure many different physical parameters depending upon how it is attached to the physical system.

Summary

In this chapter we have seen how the parameters of level, height, mass, force (or weight), and volume are related, and looked at a number of devices commonly used to measure them.

There are many other devices, and adaptations of the ones mentioned, in use. Often more than one type of device, or devices in groups, may be used within a system.

> **Questions for further discussion**
>
> 1. Why would using a capacitance probe for measuring the depth of aviation fuel in an aircraft fuel tank be more suitable than using (a) a bubbler level gauge, or (b) a float level gauge?
> 2. Consider the ultrasonic level indicator configuration shown in Figure 4.10. What uses other than measuring the depth of a liquid could this technique easily be adapted for?
> 3. Why may capacitance probes be considered unsuitable to measure the depths of certain liquids such as (a) mercury and (b) copper sulphate solution.
> 4. How could a spring balance be adapted to measure a rotational force, such as the torque produced by a motor?

5 Measurement of Pressure

Both pressure and stress have a similar basic definition. This is that they are a measure of a force acting on an area. Hence they are both measured in the same units of newtons per square metre ($N.m^{-2}$). Pressure is a fairly general term, and it is really a type of stress. When discussing the force produced by a fluid, such as air or a flowing liquid, it is usually referred to as pressure. The force caused by or acting on a solid object is usually referred to as stress.

Because of the weight of air, everything on the surface of the Earth is under pressure. This is known as atmospheric pressure. The standard value of atmospheric pressure at sea level is $1.01325 \times 10^5\ N.m^{-2}$. However, in reality atmospheric pressure is constantly varying. Next time you see or hear a weather forecast, listen for references to high and low pressure. These are different atmospheric pressures. It also changes with altitude. The higher up you are, the less air is above you, so the atmospheric pressure is less.

Because of the variations in atmospheric pressure, it is not always convenient to measure pressures absolutely (absolute pressure is measured with respect to zero pressure). Absolute pressure measurement requires an accurate measurement of atmospheric pressure. A term commonly used in pressure measurement is gauge pressure, which uses atmospheric pressure as its zero point. Absolute pressure is then the sum of gauge pressure and atmospheric pressure.

Key fact

Gauge pressure is the pressure in addition to atmospheric pressure. When something is at atmospheric pressure its gauge pressure will be zero.

The SI unit of pressure is the pascal (Pa) which is equal to one newton per square metre. You may meet pressure expressed in other, non-SI units, such as the bar, millimetres of mercury (the torr), or the atmosphere. These are often more convenient to use than the pascal, whose numbers tend to become rather large. One bar is equivalent to 100000 Pa and one millimetre of mercury is equivalent to 133.32 Pa. One atmosphere is equivalent to the standard value of atmospheric pressure quoted previously.

There are several types of pressure sensing device. When measuring the pressure of liquids the values are usually significantly above atmospheric pressure. When measuring the pressure of gases the pressures can be above or below atmospheric pressure. Because of these different ranges, and sometimes because of the different behaviour of liquids and gases, many pressure gauges can only measure the pressures of one or the other.

Most pressure sensors require calibration. For high accuracy sensors, a manufacturer will take special care and will supply a calibration certificate for each individual sensor. As the calibration may change over time, repeat calibrations may well be needed from time to time. For the highest accuracy, many pressure sensors are calibrated before each use.

Liquid manometers

Technically a manometer is any device used to measure pressure. However, the word manometer is commonly used to mean a pressure sensor which detects pressure change by means of liquid in a tube.

Manometers are differential pressure sensors. A differential pressure sensor measures the difference between a pressure being applied to it and a reference pressure (often atmospheric pressure).

Key fact

Differential pressure is a comparison of one pressure to another.

The U-tube manometer

The U-tube manometer is somewhat self-descriptive. In its basic form it consists of a clear glass or plastic tube shaped into the form of a 'U'. The tube is partially filled with a liquid, such as water, alcohol, or mercury (although for safety reasons mercury is no longer commonly used). The lower the density of the liquid, the higher the sensitivity of the manometer.

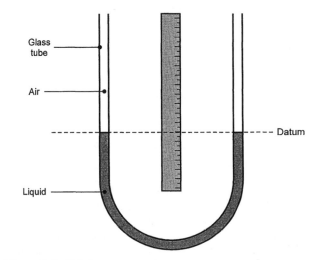

Figure 5.1 U-tube manometer

Figure 5.1 shows a basic U-tube manometer. Both ends of the tube are open, and atmospheric pressure acts equally on the liquid through each end. Therefore the height of the liquid on each side of the U (in each limb) is equal.

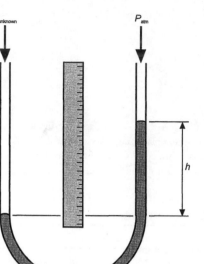

Figure 5.2 U-tube manometer with an unknown pressure applied to one limb

Figure 5.2 shows the U-tube manometer with an unknown pressure $P_{unknown}$ applied to one limb. The other limb of the tube is left as it was, that is, atmospheric pressure P_{atm} is maintained at its open end. The unknown pressure acts on the liquid in the tube, forcing it down the limb it is applied to. Because the liquid is incompressible, it rises up the other limb. Hence the height of the liquid on either side of the tube is no longer equal. The difference between the height of the liquid in each limb, h, is proportional to the difference between the unknown pressure and atmospheric pressure.

In a U-tube manometer, the difference between the unknown pressure and atmospheric pressure is the gauge pressure. In a U-tube manometer,

$$P_{gauge} = P_{unknown} - P_{atm} = \rho g h$$

where:

- ρ is the density of the liquid in the tube, in kg.m^{-3}
- g is the acceleration due to gravity, that is, 9.81 m.s^{-2}
- h is difference between the heights of the liquid in each limb, in m

If we know atmospheric pressure, or can accept the errors which occur using the standard value, then we can find $P_{unknown}$ by rearranging the equation,

$$P_{unknown} = P_{atm} + \rho g h$$

Figure 5.3 is an example calculation using a basic U-tube manometer to determine the value of an unknown pressure.

The accuracy of U-tube manometers is largely dependant on the skill of the reader in judging the difference in height of the liquids in each limb. So that this can be made easier and the height difference h can be read from one limb, a half-scale is sometimes used. Because the increase in the fluid level in one limb is directly proportional to the decrease in level in the other limb, every 1 mm change in level in one limb represents a change in h of 2 mm. A half-scale will have, say, every 5 mm labelled on one limb as 10 mm, representing the change in h.

The U-tube manometer is not in wide use in industry, although it is sometimes used to calibrate other instruments. It is mainly used in laboratories for experimental work and demonstration purposes. It can be used to measure the pressure of flowing liquids as well as gases, but cannot be used remotely. If pressures fluctuate rapidly its response may be poor and reading difficult.

Figure 5.4 shows manometers used in a piece of experimental apparatus for measuring flow of a liquid. As we shall see in Chapter 7, the pressure of a liquid in a pipe relates to its flow rate. Here, a multi-limb manometer is used, which is effectively a series of inverted U-tube manometers.

Problem

A U-tube manometer partially filled with water, has an unknown pressure applied to the end of one limb. The other end of the limb is open to atmospheric pressure. The difference between the height of the liquid in each limb is measured as 20 mm. Assuming the density of water is 1000 kg.m^{-3}, and the acceleration due to gravity is 9.81 m.s^{-2} determine the gauge pressure of the manometer.

Assuming the atmospheric pressure acting on the manometer is 1.01325×10^5 N.m^{-2}, determine the unknown pressure.

Solution

From the question, we know that h = 20 mm = 0.02 m, ρ_{water} = 1000 kg.m^{-3}, g = 9.81 m.s^{-2}, and P_{atm} = 1.01325×10^5 N.m^{-2}. To determine the gauge pressure of the manometer,

$$P_{gauge} = \rho g h$$

$$= 1000 \times 9.81 \times 0.02$$

$$P_{gauge} = 196.2 \, Pa$$

To determine the unknown pressure,

$$P_{gauge} = P_{unknown} - P_{atm}$$

So,

$$P_{unknown} = P_{atm} + P_{gauge}$$

$$= 101325 + 196.2$$

$$\mathbf{P_{unknown} = 101521.2 \, Pa \; or \; 101.5212 \, kPa}$$

Figure 5.3 Example calculation

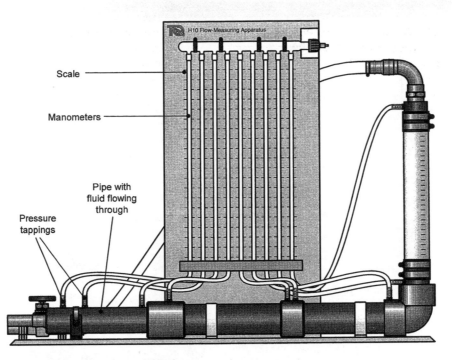

Figure 5.4   Experimental apparatus using a multi-limb manometer

The inclined tube manometer

The inclined tube manometer is a differential pressure sensor more sensitive than the U-tube manometer. Hence it is more suitable for use with smaller pressure measurements or where greater accuracy is required. Figure 5.5 shows its basic design.

One limb of the inclined tube manometer forms into a reservoir. The other limb of the manometer is inclined at a known angle θ. The inclined limb is made from a transparent material such as glass or plastic. The reservoir is usually made of plastic, but does not need to be transparent.

The surface area of the fluid in the reservoir A_1 is much larger than the surface area of the fluid in the inclined limb A_2. Both limbs are open ended and so subject to atmospheric pressure. If an unknown pressure $P_{unknown}$ is applied to the reservoir limb, the change in height h_1 will be relatively small compared to the change in height in the inclined limb h_2.

The gauge pressure is given by the equation

$$P_{gauge} \;=\; P_{unknown} - P_{atm} \;=\; \rho g d\!\left(\frac{A_2}{A_1} + \sin\theta\right)$$

where:

- ρ is the density of the liquid in the tube, in kg m^{-3}
- g is the acceleration due to gravity, that is, 9.81 m.s^{-2}
- d is the distance the liquid has moved along the inclined limb, in m
- A_2 is the cross-sectional area of the reservoir, in m^2
- A_1 is the cross-sectional area of the liquid in the inclined limb
- θ is the angle of the inclined limb from the horizontal

Because A_1 is much larger than A_2, the ratio A_1/A_2 can be considered negligible. Therefore

$$P_{gauge} \;=\; P_{unknown} - P_{atm} \;=\; \rho g d \sin\theta$$

If we know the atmospheric pressure, or can accept the errors which occur using the standard value, then we can find $P_{unknown}$ by rearranging the equation,

$$P_{unknown} \;=\; P_{atm} + \rho g d \sin\theta$$

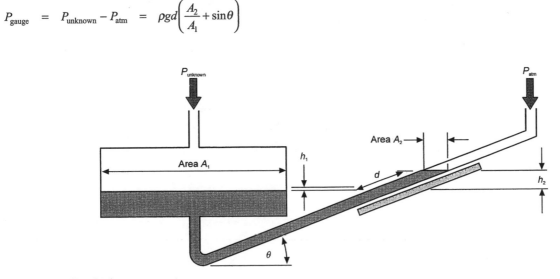

Figure 5.5   Inclined tube manometer

Front view

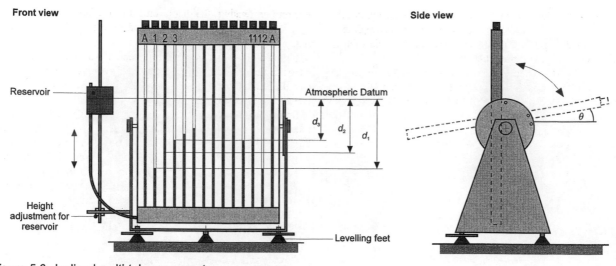

Reservoir

Atmospheric Datum

d_3 d_2 d_1

Height adjustment for reservoir

Levelling feet

Side view

θ

Figure 5.6 Inclined multi-tube manometer

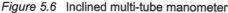

Problem	**Solution**
An inclined tube manometer containing alcohol, has an unknown pressure applied to the end of the reservoir limb. The end of the inclined limb is open and subject only to atmospheric pressure. It is inclined at an angle of 30° from the horizontal. The distance the liquid moves along the limb when the unknown pressure is applied is 25 mm. Assuming the density of the alcohol is 800 kg.m^{-3}, the acceleration due to gravity is 9.81 m.s^{-2}, and the atmospheric pressure acting on the manometer is 1.01325×10^5 N.m^{-2}, determine the unknown pressure.	From the question, we know that d = 25 mm = 0.025 m, $\rho_{alcohol}$ = 800 kg.m^{-3}, g = 9.81 m.s^{-2}, and P_{atm} = 1.01325×10^5 N.m^{-2}

To determine the unknown pressure,

$$P_{unknown} = P_{atm} + \rho g d \sin\theta$$

$$= 1.01325 \times 10^5 + (800 \times 9.81 \times 0.025\sin30°)$$

$$= 1.01325 \times 10^5 + 98.1$$

$$P_{unknown} = \textbf{101423 Pa, or 101.423 kPa}$$

Figure 5.7 Example calculation

Figure 5.6 shows a multi-limb inclined tube manometer used with experimental apparatus investigating air flow. Notice that the angle of inclination can be varied. This allows adjustment of the range and sensitivity to suit the pressure being measured.

Figure 5.7 is an example calculation using an inclined tube manometer to determine the value of an unknown pressure.

The accuracy of inclined tube manometers relies less on the skill of the reader than U-tube manometers. They are more sensitive, but unless the inclined limb is relatively long they cannot be used over as wide a range of pressures.

Inclined tube manometers are used where higher sensitivity than a U-tube manometer is required. It cannot be read remotely, and it is usually used with gases.

Elastic pressure sensors

Elastic pressure sensors are so called because something flexes, stretches, or temporarily deforms when a pressure is applied.

Key fact
Elastic pressure sensors initially convert pressure into a displacement.

This allows displacement sensors to be used to condition the output signal from the pressure sensor. Some pressure sensors are

referred to by the method they use to measure this displacement, such as piezoelectric and capacitive pressure sensors. Where electronic displacement sensors are used, the method of detecting pressure change is usually by means of a diaphragm. Elastic pressure sensors measure pressure differentially.

Bourdon tube pressure gauge

The Bourdon tube pressure gauge, named after Eugène Bourdon, is probably the most popular pressure sensor. Basic Bourdon tubes are made from metal alloys such as stainless steel or brass. They consist of a tube of elliptical or oval cross-section, sealed at one end. Their are various shapes of Bourdon tube, including helical, spiral and twisted. A common design is the C-shape, as shown in Figure 5.8.

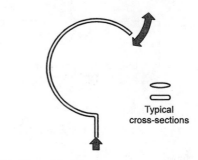

Typical cross-sections

Figure 5.8 Bourdon tube

Here the tube is at atmospheric pressure. When increased pressure is applied to the open end, it deflects outwards (tries to straighten) in proportion to the pressure inside the tube (the outside of the tube remains at atmospheric pressure).

As the pressure is decreased, the tube starts to return to its atmospheric pressure position. The amount by which the tube moves in relation to the pressure applied to it depends on factors including its material, shape, thickness, and length. Compared to other elastic pressure sensors the deflection produced by Bourdon tubes is large.

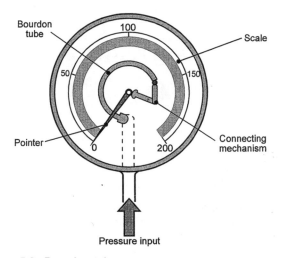

Figure 5.9 Bourdon tube pressure gauge

The Bourdon tube pressure gauge, shown in Figure 5.9, consists of a Bourdon tube connected to a pointer. The pointer moves over a calibrated scale. When pressure is applied, the movement of the tube is fairly small, so to increase the movement of the pointer it is mechanically amplified. This is usually by a connecting mechanism consisting of a lever, quadrant and pinion arrangement.

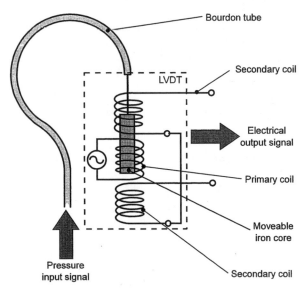

Figure 5.10 Bourdon tube and LVDT combination

Bourdon tubes need some form of compensation for temperature, as temperature changes affect their accuracy.

For remote sensing, the displacement of the Bourdon tube caused by pressure changes can be detected by a suitable displacement sensor.

Figure 5.10 shows an LVDT connected to a Bourdon tube. This converts the displacement at the end of the tube into an electrical signal. This signal can then be displayed or recorded on an electrical device calibrated in terms of pressure.

Some designs of Bourdon tube pressure gauge tend to be fairly inexpensive because they are mass produced (which reduces costs). They are suitable for use with both liquids and gases, are used in a wide variety of applications, both industrial and domestic. Applications range from tyre pressure gauges, measuring the pressure in pneumatically controlled tools and machines, to pipeline pressure in chemical plants.

Bellows

Bellows are differential pressure sensing devices mainly used in low pressure ranges of about 0 to 1000 pascals.

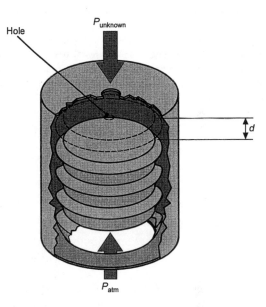

Figure 5.11 Bellows

Figure 5.11 shows a set of metallic bellows, held inside a protective casing. The bellows are made of a thin copper alloy tube pressed into a corrugated shape. This is sealed at one end, with a small hole at the other end. When pressure is applied via the hole, the bellows expand a distance d. This displacement can be calibrated in terms of pressure.

The pressure applied to the bellows P_{unknown} is given by the equation

$$P_{\text{unknown}} = \frac{d}{A}\lambda$$

where:

- d is the distance moved by the bellows in m
- A is the cross sectional area of the bellows in m^2
- λ is the stiffness of the bellows in $N.m^{-1}$

Figure 5.12 shows a single chamber 'bellow'. It consists of two circular metal diaphragms connected back to back. These devices, commonly referred to as capsules, give smaller displacements than multi-chamber bellows.

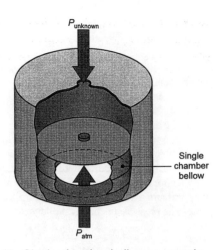

Figure 5.12 Single chamber bellow or capsule

Bellows usually produce small displacements. These need to be amplified, and displacement sensors such as LVDTs and potentiometers are often used to convert the displacement into an electrical signal. Figure 5.13 shows a bellows sensor used with an LVDT. The LVDT produces an electrical output signal directly related to the displacement input signal. This can then be displayed or recorded on a device calibrated in terms of pressure.

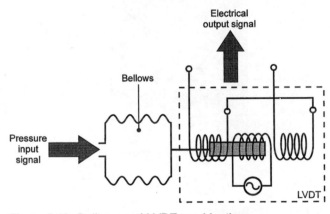

Figure 5.13 Bellows and LVDT combination

Bellows should not be used in an environment where they may be subjected to vibration or shock. Their accuracy is also affected by temperature changes.

An example of the use of bellows is in control applications, for closing valves in a pipe when a critical pressure is reached.

Capacitive pressure sensors

Capacitive pressure sensors use the electrical property of capacitance (which we met in Chapter 3), to measure the displacement of a diaphragm. The diaphragm is an elastic pressure sensor displaced in proportion to changes in pressure. It acts as one plate of a capacitor.

The diaphragm consists of a thin plate of metal alloy, such as stainless steel or brass. It is circular and fixed continuously around its edges to a cylinder. Figure 5.14 shows how it displaces when a pressure is applied to it. It is a differential pressure sensor, because it is subject to atmospheric pressure. The amount by which the diaphragm moves with respect to the unknown pressure applied to it depends on its shape and construction, size, thickness, and material. Therefore the precise relationship

between pressure and displacement varies with each design. However, the displacement is small, rarely more than a few millimetres, often fractions of a millimetre. Because of this, these devices are often individually calibrated (against an accurate inclined tube manometer, for example).

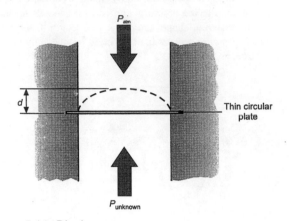

Figure 5.14 Diaphragm

Because the displacement of the diaphragm is small, a displacement sensing device with high resolution, sensitivity and accuracy is required. Also, a non-contact device is required as any mechanical resistance will reduce the displacement even more. Using the diaphragm as one plate of a capacitor is a non-contact displacement sensing method, providing an electrical output with infinite resolution particularly suited to small displacements.

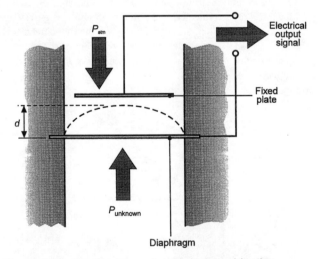

Figure 5.15 Diaphragm and capacitor combination

Figure 5.15 shows a diaphragm and capacitor combination. One plate of the capacitor is fixed, the other plate is the diaphragm. The dielectric is usually air. Notice that it does not use the variable area technique like the capacitive displacement sensor discussed in Chapter 3. Instead, the distance between the plates is varied. When there is a change in pressure, the diaphragm deflects in proportion to it, and this causes the distance d between the plates of the capacitor to change. The capacitance c of the capacitor is (approximately) inversely proportional to the distance d between the plates. That is,

$$c \propto \frac{1}{d}$$

An electrical signal is connected to the two plates of the capacitor. The change in value of the capacitance causes this electrical signal to vary. This is then conditioned and displayed on a device calibrated in terms of pressure.

There are techniques other than change in capacitance used to measure the displacement of a diaphragm. For example, another design of pressure sensor uses a diaphragm made from silicon. Semiconductor strain gauges are diffused into the diaphragm. They are arranged in a form similar to the Wheatstone bridge. A stable electrical output is produced, and the arrangement of the strain gauges compensates for temperature changes. However, these types of pressure sensor are relatively expensive.

Because of the sensitivity of diaphragm based pressure sensors, they are used to measure small changes in pressure. An example application is measuring small changes in flow rate in liquid or gas pipelines.

Piezoelectric pressure sensors

Figure 5.16 shows the principle of a piezoelectric pressure sensor. These sensors are similar to capacitive pressure sensors in that they detect pressure changes by the displacement of a thin metal or semiconductor diaphragm. We discussed the piezoelectric effect in Chapter 3. In a pressure sensor using this effect, the diaphragm causes a strain on the piezoelectric crystal when flexing due to pressure changes. The electric charges of opposite polarity appearing on the faces of the crystal are proportional to this strain.

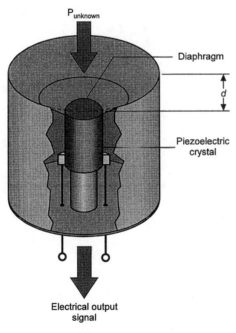

Figure 5.16 Piezoelectric pressure sensor

The piezoelectric crystal is usually quartz. This type of pressure sensor often incorporates signal conditioning circuitry in a sealed unit, using integrated circuit technology.

Piezoelectric pressure sensors operate at high temperatures and can be made small in size. Their main advantage is that they have a fast response and fairly wide operating range. Hence they can be used in applications such as measuring the pressure in a gun barrel when it is fired. They have very high sensitivity and also good accuracy, repeatability and low hysteresis.

Piezoresistive and strain gauge pressure sensors

A pressure sensor similar to the piezoelectric pressure sensor is the piezoresistive pressure sensor. These devices use silicon diaphragms, which form part of a semiconductor integrated circuit chip.

> **Key fact**
>
> Piezoresistivity is a strain dependent resistivity in single crystal semiconductors.

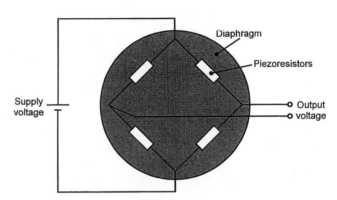

Figure 5.17 Piezoresistive diaphragm

Four piezoresistors on the diaphragm form a Wheatstone bridge. This is shown in Figure 5.17.

When pressure is applied to the diaphragm it causes a strain in the resistors. The resistance of the piezoresistors changes in proportion to this strain, and hence the change in pressure.

Figure 5.18 is a typical specification of a piezoresistive pressure transducer.

Piezoresistive pressure sensor
Range
0–50 kPa
Full-scale output
50 mV
Accuracy
±0.2%
Sensitivity
1 mV per kPa
Maximum overpressure
140 kPa
Response time
<1 ms
Excitation
10 V
Repeatability and hysteresis
±0.01% FSO

Figure 5.18 Piezoresistive pressure sensor specification

The range of piezoresistive pressure sensors is normally low, typically at a maximum of about 0 to 200 kPa. They are used where good repeatability, high accuracy, low hysteresis and long term stability are required. A typical application is in sensing the pressure at the bottom of a water tank. This pressure is directly related to the depth of water, and so the pressure sensor forms part of a level measurement system.

For measuring higher pressure ranges, instead of piezoresistors, bonded resistance strain gauges may be used. These are based on a similar principle to the piezoresistive strain gauge, in that the resistors are bonded to a diaphragm and formed into a Wheatstone bridge. They can measure pressures up to about 25000 kPa.

Barometers

A barometer is a pressure sensor specifically used to measure atmospheric pressure. Because of this they have to be sensitive, and measure absolute pressure. They are mainly used for meteorological purposes. High atmospheric pressure is usually associated with fine weather while low pressure usually predicts poor weather.

Barometers have existed for many years, and there are two basic types.

Liquid in glass barometer

This barometer is a type of liquid manometer. It consists of a thin glass tube containing a liquid. The liquid is usually mercury.

Figure 5.19 shows the basic design of a liquid in glass barometer. It consists of a U-shaped manometer fixed to a backing plate. One limb forms a reservoir, and is open to atmospheric pressure via a small hole. The other limb is a thin tube, containing a volume of liquid small in comparison to that contained in the reservoir. Like the inclined tube manometer, this allows the pressure to be read against a scale on one limb. The end of this limb contains a vacuum and is sealed.

Key fact

A pure vacuum is a space totally devoid of any matter, which therefore has no pressure. On Earth a pure vacuum is impossible to achieve. However, the term vacuum is applied to something having a pressure lower than atmospheric pressure.

It is the vacuum which allows the barometer to measure absolute pressure. It is still measuring the difference in pressure between the two limbs. However, the pressure produced by the vacuum is lower than the local air pressure, and constant. Therefore, the pressure acting on the limb containing the vacuum can be considered negligible. The only significant pressure acting on the liquid in the barometer is the atmospheric pressure. Hence the height of the liquid in the tube is proportional to atmospheric pressure.

The level of the reservoir may be adjusted on some liquid in glass barometers. This allows it to be matched to the zero of the scale before readings are taken thus improving accuracy.

There are several designs of liquid in glass barometer. They are accurate but can be rather cumbersome. They also tend to be fragile, and their accuracy is significantly affected by temperature variations. They are usually in a fixed position and are often used for calibration of other pressure sensors or as a datum in laboratories.

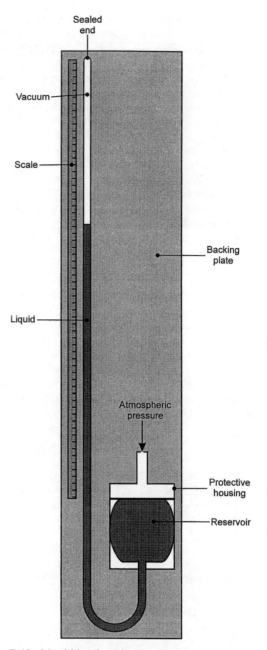

Figure 5.19 Liquid in glass barometer

Aneroid barometer

The word aneroid simply means something which does not contain liquid, and so distinguishes this type of barometer from the liquid in glass type.

The aneroid barometer, shown in Figure 5.20, senses atmospheric pressure changes by means of a sealed metal capsule. One or two faces of the capsule are diaphragms. The capsule contains a partial vacuum. Because the pressure inside the capsule is lower than atmospheric pressure and constant, it is sensitive to changes in atmospheric pressure. If atmospheric pressure increases, the diaphragms move and the capsule flattens. If atmospheric pressure decreases, the capsule expands. This displacement is detected by a bar pressed on to the capsule by a strong spring. This is mechanically amplified by gears and levers and transmitted to a pointer on a calibrated scale.

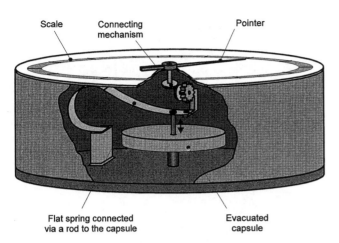

Figure 5.20 Aneroid barometer

The aneroid barometer is simpler to read and takes up less space than the liquid in glass barometer. It is not affected by temperature changes as signiffcantly, but is less accurate mainly because of its mechanical parts. In fact, aneroid barometers are usually calibrated against liquid in glass barometers.

Aneroid barometers are commonly used for domestic purposes. They are likely to be calibrated in non-SI units, such as millimetres of mercury or bar. Some versions are calibrated with respect to likely weather conditions!

If electronic measurement is required, the movement of the diaphragm can be detected in terms of capacitance as discussed earlier in this chapter. This increases accuracy and sensitivity. Because atmospheric pressure changes with altitude, capacitive aneroid barometers sometimes form the basis of altimeters in aircraft.

Summary

In this chapter we have looked at pressure, and devices used to measure it. Pressure can be measured differentially, as gauge pressure, or absolutely. Pressure sensors can be loosely categorised as manometers, elastic pressure sensors, and barometers. There are several other types of more specialised pressure sensor available, such as those to measure 'high vacuums'. This chapter will have given you a basic understanding of pressure and the devices commonly available to sense it.

Pressure itself can be used to determine other parameters, such as the depth of a liquid, fluid flow rate, or height of an aircraft. The choice of pressure sensor depends on many factors such as the range and type of pressure to be measured, and the accuracy of measurement required.

Questions for further discussion

1. In a simple U-tube manometer, how does the internal diameter of the tube affect its operation?
2. What problems may arise if a mechanical aneroid barometer was used for altitude indication in an aircraft?
3. Why, in the past, do you think mercury was the most popular liquid to use in manometers and liquid in glass barometers?
4. Consider the piezorcsistive pressure sensor specification shown in Figure 5.18. What practical applications would this sensor be particularly suited for?
5. Even though the displacement of diaphragms is very small, why do you think they are more commonly used in pressure sensors than bellows?

---

6 Temperature Measurement

Temperature is the degree of hotness of one body, substance, or medium compared with another. When measuring temperature, we usually compare this degree of hotness to a fixed reference point, using temperature scales. The Kelvin thermodynamic scale uses absolute zero as its reference point. The Celsius (which used to be called centigrade) scale uses a point of reference based on the freezing-point of water (0 °C) and the boiling point of water (100 °C).

Key fact

Absolute zero is the lowest temperature any substance can reach. Molecules of a substance contain no heat energy at absolute zero, which is 0 K or about −273.15 °C.

Temperature measurement is important because, at different temperatures, substances have different physical properties and behave in different ways. For example, the temperature of a substance will affect its electrical properties, whether it is solid, liquid, or a gas, and it will also affect its volume. Small changes in body temperature can show whether a person or animal is ill.

In this chapter we will look at devices which sense temperature and temperature change, and the properties of certain substances which allow temperature measuring devices to work. Temperature measuring devices are called thermometers, sometimes referred to as pyrometers if they measure high temperatures.

There are many different types of thermometer. The main types of thermometer we shall look at are ones which measure temperature by means of:

- Liquid expansion.
- Metal expansion.
- Electrical resistance.
- Thermoelectricity.
- Heat radiation.

Liquid expansion thermometers

Liquid expansion thermometers work on the principle that certain liquids change in volume by a large amount when they change in temperature, in comparison to certain solids whose volume changes by a relatively small amount. The amount by which a liquid expands can be calibrated in terms of temperature change.

The liquid-in-glass thermometer clearly illustrates this principle, since glass expands little compared to most liquids, for the same temperature increase.

Liquid-in-glass thermometer

Figure 6.1 shows a typical liquid-in-glass thermometer. It consists of a glass tube, sealed at both ends, with a fine bore or column at its centre, containing a liquid. At the base of the column the bore opens up to form a reservoir, known as the bulb. The bore also opens out at the top to provide an expansion cavity. When the thermometer is heated the liquid expands and rises up the column. Above the liquid is a vacuum, or a gas which compresses when the liquid expands. A scale on the side of the stem is calibrated so the height of the liquid in the column is proportional to the temperature of the thermometer.

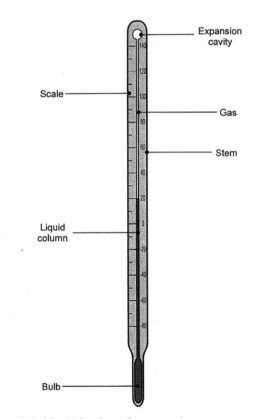

Figure 6.1 Liquid-in-glass thermometer

Commonly used liquids in this type of thermometer are mercury, alcohol, or synthetic oils. There are other liquids used in liquid-in-glass thermometers, suitable for more extreme conditions, or, because mercury is toxic, for safety reasons in case of breakage. The liquid used in the thermometer depends on the temperature range to be measured. It must not freeze or boil within the given temperature range, and it should change volume linearly with respect to changes in temperature.

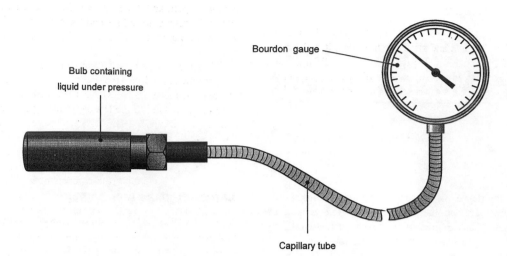

Figure 6.2 Liquid-in-metal thermometer

For example, mercury in glass thermometers have a temperature range of approximately 238 K to 783 K (−35 °C to +510 °C), whereas alcohol in glass thermometers have a temperature range of approximately 193 K to 343 K (−80 °C to +70 °C).

Liquid-in-glass thermometers are used in medical and veterinary applications for taking the body temperature of humans or animals. Another common use is in indicating air temperature for domestic and meteorological purposes.

Liquid-in-glass thermometers are inexpensive, simple to use, and reliable. However, they are fragile and need careful handling, and an environment where they are not likely to be knocked by other objects or subject to vibration. Their response to rapid temperature changes is poor and they may only be read locally. Accuracy can be good but is largely dependent on the skill of the reader. They are not suitable for surface temperature measurements. Hence their industrial applications are limited, but they are used occasionally, for example, fitted into pipelines.

Liquid-in-metal thermometer

Liquid-in-metal thermometers work on the same principle as liquid-in-glass thermometers, that is, the expansion of a liquid can be calibrated in terms of temperature change. Figure 6.2 shows a typical liquid-in-metal thermometer. It consists of a metal bulb (often made from stainless steel) containing a liquid, such as mercury or alcohol, which in some cases is under pressure. The metal bulb connects to a flexible capillary tube. Unlike the liquid-in-glass thermometer, the temperature is not read by looking at the liquid in the capillary tube. Instead, the capillary tube is connected to a Bourdon gauge (see Chapter 3), which is calibrated in terms of temperature. When the liquid expands due to an increase in temperature, the Bourdon tube straightens slightly. Gears and linkages amplify this movement to a pointer which moves over a scale to provide a direct temperature reading.

The main advantages of liquid-in-metal thermometers are that they are more robust than liquid-in-glass thermometers, and they can be read remotely (up to about 35 metres). However, errors may occur because of the temperature of the Bourdon tube or capillary tube changing. They are generally more expensive than liquid-in-glass thermometers.

Typical applications of liquid-in-metal thermometers are in chemical plants, vehicle engines, and measuring the temperature of certain liquid metals.

Metal expansion and the bimetallic strip

A bimetallic strip is a device consisting of two dissimilar metal strips of the same length, secured together by riveting, welding, or bonding, and having different expansivity (sometimes called coefficient of expansion).

> **Key fact**
>
> The expansivity of a material is the fraction of its original dimension by which the substance expands per degree rise in temperature.

The two metals which form the strip are usually an iron-nickel alloy with very small expansivity, and a metal with high expansivity, such as brass. Figure 6.3 shows a bimetallic strip.

One end of the bimetallic strip is fixed, and when there is an increase in temperature the bimetallic strip bends into a curve. This is because of the different expansivities of the two metals. Brass has the higher expansivity of the two metals and expands significantly more than the iron-nickel alloy for the same temperature rise. The brass becomes longer than iron-nickel alloy and the bimetallic strip bends so the brass is on the outside of the bend. The amount by which the free end moves is related to the temperature change.

Bimetallic strips are used in thermostats and bimetallic thermometers. They also have applications in electromechanical components such as automatic flashing units for motor vehicle direction indicators and automatic flashing units for advertising signs.

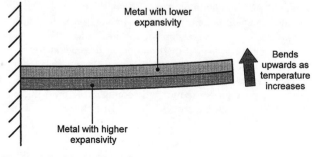

Figure 6.3 Bimetallic strip

Bimetallic thermostat

The bimetallic strip can be positioned so that when it is heated and bends, or when it cools and starts to straighten, it will connect or disconnect with a terminal in an electrical circuit. As a consequence, the electrical circuit will make or break. Thermostats use this effect to control the heat produced by a heating system.

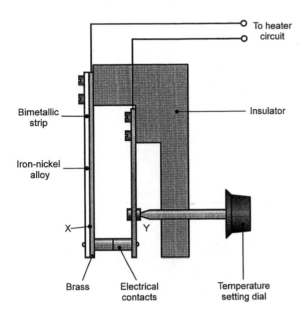

Figure 6.4 Thermostat

A thermostat is a device which keeps a system or a substance at a constant temperature. Figure 6.4 shows a typical domestic thermostat, used for controlling the temperature of a room in a house.

The heater circuit connects to the bimetallic strip (X) and the metal strip (Y). In the position shown the contacts are closed and an electric circuit switches on the heater. As the temperature rises above the set temperature, the bimetallic strip bends. Hence the

contacts open, and the electrical circuit to the heater disconnects. When the temperature falls and the contacts will eventually close, and once again the electrical circuit is made, so the heater operates.

Adjusting a screw on the temperature setting dial controls the temperature at which the contacts open. This screw presses against an insulating sleeve on the metal strip (Y). If the metal strip (Y) is moved to the left by adjusting the temperature setting control, the bimetallic strip (X) will have to bend further before the contacts open. Consequently a higher room temperature is maintained. If, by adjusting the temperature setting control, the metal strip (Y) is moved to the right, the bimetallic strip (X) will have to bend less before the contacts open. Hence a lower room temperature is maintained.

Apart from domestic temperature control, other typical applications of the thermostat switch are in electric irons, immersion heaters, aquariums, ovens, and electric fires. Next we'll take a look at the bimetallic thermometer.

Bimetallic thermometer

Figure 6.5 shows a typical bimetallic thermometer. The bimetallic strip is coiled in a helical form to increase sensitivity. A change in temperature produces a twisting of one end of the bimetallic strip relative to the other.

A spindle connected to the bimetallic strip transfers the twisting movement to a pointer over a calibrated scale. This provides a direct reading of temperature.

The bimetallic thermometer is used for measuring the temperatures of ovens, hot water pipes, and steam chambers. It is compact, robust, non-electrical, relatively inexpensive and has a useful operating range of about 238 K to 873 K (–35 °C to +600 °C) with fairly good accuracy. However, regular recalibration is necessary because of ageing effects on the bimetallic strip. Also, bimetallic thermometers are not suitable for remote use, and respond slowly to temperature changes.

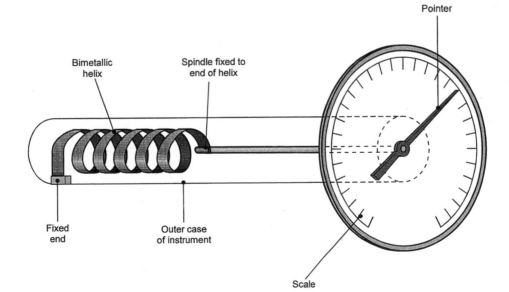

Figure 6.5 Bimetallic thermometer

Electrical resistance

Most metals increase in electrical resistance as their temperature increases. The relationship between resistance and temperature is, over a limited range, almost linear and is given by the expression:

$$R_t = R_0 (1 + \alpha t)$$

where:

- R_0 is the resistance in ohms of the conductor at a temperature of 0 °C
- R_t is the resistance in ohms of the conductor at t °C
- α is the temperature coefficient of resistance of the material

Note the expression uses a property of material known as the temperature coefficient of resistance, α. This property varies for each different material. Figure 6.6 gives some typical values of α for various metals. Figure 6.7 is an example of calculating the change in resistance of a metal when it is subjected to a change in temperature.

Metal	Temperature coefficient of resistance α
Copper	4.3×10^{-3}
Silver	3.9×10^{-3}
Iron	6.5×10^{-3}
Nickel	6.5×10^{-3}
Platinum	3.9×10^{-3}

Figure 6.6 Typical values of temperature coefficient of resistance for various materials

Electrical resistance thermometer

Electrical resistance thermometers use the relationship between the temperature of a metal and its electrical resistance.

Figure 6.8 shows a typical design of electrical resistance thermometer (although there are many other designs, depending on the intended application). Wire is wound over a ceramic coated tube to form a coil, coated in ceramic and fixed inside a protective casing. This arrangement forms a temperature probe. The ends of the coil of wire are connected to one arm of a Wheatstone bridge.

The probe is placed in the substance whose temperature is to be measured and after a response time of several seconds, the temperature of the substance is displayed on the meter.

The leads that connect the coil of wire to the Wheatstone bridge also change in resistance with changes in temperature, which introduces errors into the measurement process. To counteract these errors compensating leads are connected in an adjacent arm of the bridge, as shown. The imbalance of the bridge is directly proportional to the change in the probe resistance – the indicator can thus be calibrated as a temperature scale.

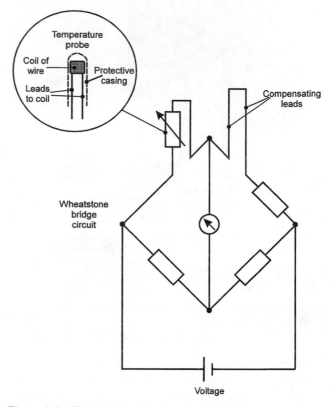

Figure 6.8 Electrical resistance thermometer

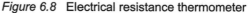

Problem
A length of platinum wire has a resistance of 110 ohms at the temperature of 0 °C. Calculate the resistance of the length of platinum when the temperature is raised to 25 °C, and so determine the change in resistance.

Solution
Now $R_0 = 110\ \Omega$, $\alpha = 3.9 \times 10^{-3}$, $t = 25$ °C

$$R_t = R_0 (1 + \alpha t)$$

$$R_{25} = 110 \times \left(1 + [3.9 \times 10^{-3} \times 25]\right)$$

$$R_{25} = 110 \times (1 + 0.0975) = 110 \times 1.0975$$

$$R_{25} = 120.725$$

Resistance of the length of platinum at 25 °C 120.725 Ω. The change in resistance is ΔR.

$$\Delta R = R_t - R_0$$

$$\Delta R = 120.725\ \Omega - 110\ \Omega = 10.725\ \Omega$$

Change in resistance, ΔR, is **10.725 Ω**

Figure 6.7 Example calculation

Nickel, copper, and some other types of metal have been used in electrical resistance thermometers though nowadays platinum is by far the most commonly used. Despite being expensive, platinum has the advantage of often being the reference material for international standards. The relationship between temperature and resistance is linear over a wide range for platinum. It also has a high melting point and is thus useful for high temperature measurement.

Electrical resistance thermometers are available in many forms, and so have a wide range of applications. They can measure the temperature of gases and liquids, the surface temperature of most solids and the internal temperature of some soft solids. They are stable and can withstand harsh environments, and are used in the chemical industry for measuring the temperature of corrosive liquids and slurries. The food industry uses this type of thermometer to measure temperatures of foodstuffs such as meat.

Electrical resistance thermometers are accurate but have a slow response time, are fairly large and fragile, and are also expensive.

Thermistor

The coil of wire in the metal electrical resistance thermometer has a disadvantage in that the changes in resistance are comparatively small, typically 5 milliohms per degree celsius. Semiconductor resistors, known as a thermistors (an abbreviation of thermal resistor) are also used for temperature measurement. Thermistors use the same principle as metal electrical resistance thermometers, that is, that resistance changes with temperature. Instead of metal, however, thermistors use semiconductors.

Semiconductors provide much larger changes in resistance for the same temperature change. They are made from mixtures of 'rare earth' metal oxides such as manganese, nickel, chromium and cobalt (not germanium or silicon, which are usually associated with semiconductor devices), mixed with finely divided copper. The resistance of these materials is very sensitive to temperature change.

Thermistors are available in many forms, including discs, beads and rods shown in Figure 6.9.

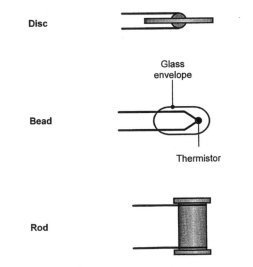

Figure 6.9 Various forms of thermistors

The resistance of a thermistor normally decreases with an increase in temperature. The relationship between resistance and temperature is exponential, as opposed to the almost linear relationship of the metal electrical resistance thermometer. The resistance of a thermistor is given by:

$$R_t = R_0 \exp \beta \left(\frac{1}{T} - \frac{1}{T_0} \right)$$

where:

- R_t is the resistance of the thermistor at T kelvin (K)
- T is absolute temperature in kelvin
- R_0 is the resistance of the thermistor at reference temperature T_0 kelvin
- β is the thermistor constant in kelvin

Note the expression uses a property of thermistors known as the thermistor constant β. The thermistor constant varies between devices, and is dependent on the material and manufacturing techniques. Typically, β ranges from 3000 K to 5000 K and R_0 is 2000 Ω.

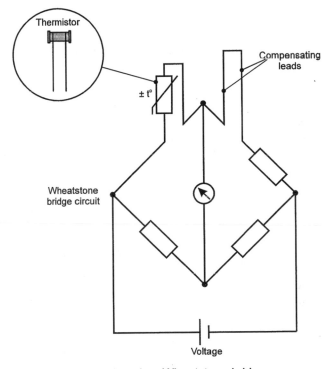

Figure 6.10 Thermistor in a Wheatstone bridge

The thermistor can be connected into one arm of a Wheatstone bridge in a similar manner to the electrical metal resistance thermometer, as shown in Figure 6.10. Note that the thermistor has its own electrical symbol.

Self-heating occurs because of current from the supply and this causes drift. Compensation for this effect is usually provided by a second thermistor held at a constant temperature.

Thermistors can be made very small and still have a high resistance, giving quick response to temperature changes. The temperature range of thermistors is normally from approximately 173 K to 573 K (–100 °C to +300 °C), but higher and lower ranges are possible. They may be used for measuring temperatures in a small area. Because they have good repeatability and fine resolution over narrow ranges, they are often used in medical applications. They are also commonly used for monitoring temperature of electronic circuitry, and they can be easily encapsulated in solids to act as surface temperature probes.

Thermistor

Resistance
At 20 °C (approx.): 2 kΩ
At 25 °C (nominal): 1680 Ω ±20%
At 200 °C (approx.): 37 Ω

Minimum operating resistance
37 Ω

Characteristic temperature
25 °C to 85 °C (nominal): 3050 K ±5%.

Dimensions
Length: 76.2 3.2 mm
Diameter: 4 mm

Self-heating effect in air
The temperature of the thermistor element rises by 1 °C for each 1.3 mW of power dissipated

Maximum bead temperature
200 °C (Provided that the maximum power dissipation is not exceeded)

Maximum continuous power dissipation in free air at 20 °C
Averaged over any 20 ms period: 230 mW

Nominal thermal cooling time constant T in free air from the self-heated state
20 s

Mass (nominal)
1.8 g

Figure 6.11 Thermistor specification

Thermoelectricity

Thermoelectricity is the relationship between the temperature of a substance and electrical energy. Under some circumstances, electrical energy and heat (thermal) energy may convert from one to the other. If the change in electrical energy caused by the conversion of thermal energy can be measured, it can be related to the temperature of the substance.

When a pair of different metals are formed in a loop with two junctions at different temperatures, a current will flow whose value is related to temperature. This is called the Seebeck effect.

Key fact

The Seebeck effect is where an e.m.f. is set up in a circuit in which there are junctions between different substances, the junctions being at different temperatures.

The Seebeck effect may occur, for example, where two different types of metals meet as part of a circuit in a circuit board, and have junctions at different temperatures. Here, even though the resultant e.m.f. may be small, it could be a nuisance and need consideration when designing the circuit. However, the Seebeck effect is useful in that it is the basis of using thermoelectricity for measuring temperature.

Consider the diagram in Figure 6.12. Metal X and metal Y are dissimilar, and junction 1 and junction 2 are at different temperatures, T_1 and T_2. Because of the Seebeck effect, small e.m.f.'s are generated across the junctions. The algebraic sum of the two e.m.f.'s causes a current to flow. This thermoelectric effect is such that, for two given metals, at given different temperatures, the resultant e.m.f. is always the same. Hence, the resultant e.m.f. can be measured and calibrated in terms of temperature.

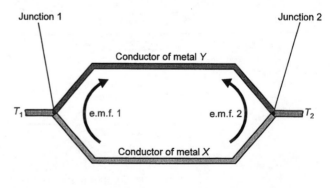

Figure 6.12 Theory of the thermocouple

If both of the junctions are at the same temperature, no e.m.f. will be generated. If the temperature of one junction starts to change, and not the other, an e.m.f. will be generated and get larger as the temperature difference between the two junctions increases. This is the basis of the thermocouple.

Thermocouple

A thermocouple has two dissimilar metals joined to form a closed-circuit. Figure 6.13 shows a typical thermocouple design. A probe or protective sheath holds one junction. This is placed in the substance whose temperature is to be measured.

If one junction is kept at a different temperature from the other then a current will flow. The magnitude and direction of the current depends upon which metals are used and the temperature of the junctions. The size of the resultant e.m.f. is small, usually in the order of millivolts. A voltmeter at the cold end of the circuit provides an output to the user, usually calibrated in temperature.

For accurate results, one junction must be maintained at a fixed temperature thus eliminating errors due to drift. The resultant e.m.f. is not affected by the size of the conductors, the areas of the junctions or how the metals are joined.

Typical metals used for thermocouple wires are rhodium, alloys of nickel and chromium, alloys of aluminium and nickel, or alloys of nickel and copper. The dissimilar metals paired with these include platinum, copper, and iron. The protective covering over the probe can be made of a number of materials, to provide strength or protection from corrosive environments.

Thermocouples are in wide use for temperature measurement, and can measure the temperature of solids, liquids, or gases. They are fairly low cost and very versatile. They have a range from 73 K to over 2273 K (–200 °C to over 2000 °C), the precise nature of which depends on the type of wires used. They respond quickly to changes in temperature, but they are generally not as accurate as devices such as resistance thermometers.

Thermocouples are used widely in industrial temperature measurement, for example in furnaces, in liquid metals, and even in nuclear reactors. They are used in medical applications such as monitoring internal temperature during operations. One reason for the popularity of thermocouples is that they can measure the temperature of objects which are very small, for example semiconductor electrical components. A typical thermocouple specification is shown in Figure 6.14.

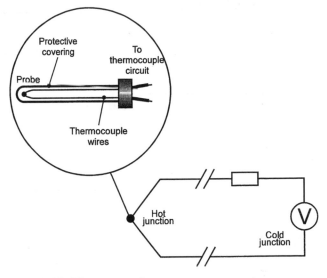

Figure 6.13 Thermocouple

Thermocouples are often used in groups. This is done to increase sensitivity, and these groups are known as thermopiles.

Thermocouple

Specifications
1 mm diameter, 150 mm long probe with standard tip and potseal PVC termination
Thermocouple, 2 mm diameter, 75 mm long probe with hypo tip and handle

Manufactured to the British Standard
Stainless-steel sheathed (unless otherwise specified)
Single hot-junction
Grounded (floating can be supplied on request) with various insulation

Probes
Hand-held with handle and leads or mounted and mineral-insulated with various terminations

Available diameters of wire/mineral-insulated probe
0.5 mm, 1 mm, 1.5 mm, 2 mm, 3 mm, 4.5 mm, 6 mm, 8 mm

Available sheath lengths
25 mm, 75 mm, 100 mm, 150 mm, 200 mm, 250 mm, 500 mm, 1000 mm

Available terminations
PVC lead, PTFE lead, handle, potseal PVC, potseal PTFE

Available tips
Naked lead, standard, mineral-insulated sheath, hypo, patch

Figure 6.14 Typical thermocouple specification

Heat radiation

All substances emit heat radiation. When a substance or body is at or above a suitable temperature, we can see it glow. This is heat energy being transferred from the substance by radiation of electromagnetic rays. If the substance changes colour or becomes brighter, it indicates that more heat energy is radiating off the substance – it is hotter.

We are not always able to see the radiation emitted by a body because the electromagnetic rays will be out of range of human eyesight, for example in the infra-red spectrum. However, bodies emit heat radiation even though we are unable to see the rays, and there are devices which can detect them.

Whether we detect electromagnetic rays with our eyes (for example from an electric light-bulb), or whether we use an instrument to detect them, the intensity of the electromagnetic radiation from a body relates to its temperature.

Different substances at different temperatures radiate energy at different rates. A body which completely absorbs any heat or light radiation falling on it is known as a black body (black bodies only exist in theory). The amount of energy given off by the surface of a particular substance is called its emissivity, ε.

> **Key fact**
>
> Emissivity, ε, is the ratio of the radiation emitted by a given surface to the radiation emitted by the surface of a black body heated to the same temperature of the surface, under the same conditions.

Emissivity is a dimensionless number between 0 and 1. A black body would have an emissivity of $\varepsilon = 1$, a typical value for polished copper is $\varepsilon = 0.3$, and for a rough, dark surface is $\varepsilon = 0.8$.

If we know the emissivity of a substance, then by measuring the electromagnetic radiation coming from it, we can measure its temperature. At a given wavelength, brightness will vary with temperature. The radiation can be measured by looking at its intensity or colour. Devices which measure temperature in this way are referred to as radiation pyrometers.

Radiation pyrometers usually measure high temperatures. A major advantage of them is that, unlike the other devices we have looked at in this chapter, they do not have to be in contact with the substance whose temperature they are measuring.

Disappearing filament optical pyrometer

The disappearing filament optical pyrometer uses the electromagnetic radiation which can be seen by the human eye to measure temperature. Therefore, the body whose temperature is being measured, such as a furnace, has to be hot enough for the human eye to see it glow. This is usually hotter than about 923 K (650 °C).

The disappearing filament optical pyrometer compares the visible electromagnetic radiation given off by the hot body with the light emitted by a lamp. The lamp is calibrated so that its filament brightness corresponds to known temperatures.

Figure 6.15 shows a disappearing filament optical pyrometer. It is in the form of a telescope, which is aimed at the hot body. By looking through the eyepiece, the user sees a small section of the hot body, and the lamp filament superimposed on top of it. They can then compare the brightness of the lamp with the brightness of the hot body.

Adjusting the current to the lamp will adjust the lamp brightness. The user alters the current until the filament seems to disappear. This indicates a similar temperature to that of the body. If the filament appears too dark, it is not as hot as the body being measured. If the filament appears too bright, it is hotter than the body being measured. The current to the lamp at which the filament disappears is thus an indication of the temperature of the body.

The absorption filter enables the device to measure temperatures much higher than the temperature of the lamp filament. The wavelength filter ensures the brightness is compared at one wavelength.

Disappearing filament optical pyrometers are used for applications such as measuring the temperatures of molten metals, furnaces, and heat-treatment processes.

The accuracy of optical pyrometers is not as good as some of the other devices we have discussed. However, they can be used to measure higher temperatures than many devices, or where contact with the substance whose temperature is to be measured is not possible. Disappearing filament optical pyrometers cannot be used for remote measurement.

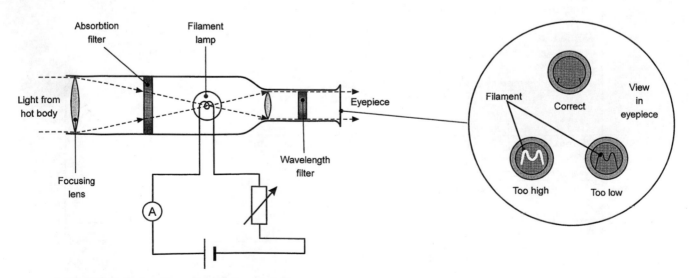

Figure 6.15 Disappearing filament optical pyrometer

Infra-red pyrometer

Infra-red pyrometers allow remote reading and rely less on the judgement of the operator than disappearing filament optical pyrometers. They use the infra-red electromagnetic rays emitted by a body and measure the intensity of them by a device such as a thermocouple or thermopile.

Figure 6.16 shows a typical infra-red pyrometer. The operator focuses the telescope arrangement on the hot body. A dichroic mirror reflects a particular band of electromagnetic radiation, such as infra-red, but allows all others through, allowing the user to focus on the hot body. The infra-red from the hot body reflected by the mirror is focused on to a thermocouple or thermopile. The output from the thermocouple is calibrated in temperature.

Infra-red pyrometers can measure a wider range of temperatures than disappearing filament optical pyrometers. This is because they can detect electromagnetic rays which are not visible to humans. The thermocouple circuit allows remote reading, but also adds the limitations and inaccuracies of the thermocouple to the system.

Infra-red pyrometers are used in the same sort of applications as disappearing filament optical pyrometers, but where remote reading, or a lower or wider range of temperature measurement is required.

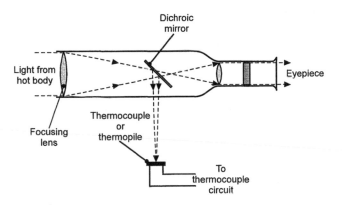

Figure 6.16 Infra-red pyrometer

Summary

In this chapter we have looked at temperature, and common devices and techniques used to measure it. Heat is a form of energy which affects the behaviour of materials in several ways, and it is these changes in behaviour which we use, by various methods, to measure the temperature of a substance.

All the devices we have looked at are available in varying designs and calibration ranges. The choice of device depends on its limitations, and the range and type of temperature measurement required.

There are other devices, techniques, and adaptations of the ones we have discussed, available to measure temperature. This chapter will have provided you with a good insight into the subject of temperature measurement.

Questions for further discussion

1. What are the limitations of mercury-in-glass thermometers compared to mercury-in-metal?
2. Under what circumstances would copper be a more appropriate choice than platinum for the wire probe in a resistance thermometer?
3. What factors need to be considered when specifying a device to measure the temperature of water in the hot water tank of a domestic central heating system? What device or devices would be suitable?
4. What problems could you foresee in using a disappearing filament optical pyrometer to measure the temperature of a metal heat treatment furnace? How could these be overcome?
5. Consider temperature measurement in a highly hazardous environment, such as a nuclear reactor. What devices could be used here? Justify your answers and mention any special precautions you may make.

---

7 Flow Measurement

Flow is the continuous movement of fluid as a stream. It occurs in many forms, from the water that flows in streams and rivers, the oil and gas that flows though pipelines, to the air that flows in and out of our lungs when we breathe. To be able to understand and use the nature of flow as best we can, it is important to be able to quantify and measure flow appropriately.

In this chapter we'll look at the three main areas of flow measurement, which are measurement of volume, measurement of mass or weight, and measurement of velocity. Because the nature of flow can be complex (like the flow of air into our lungs), in this book we'll consider simple, steady flows such as that represented by the streamlines in Figure 7.1.

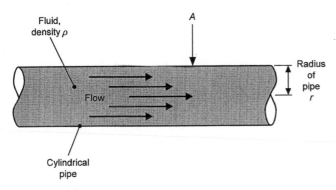

Figure 7.1 Flow through a pipe

Consider the section of cylindrical pipeline shown in Figure 7.1. Assuming the fluid is flowing steadily and there is no air in the pipe, we can quantify the flow in terms of volume, mass, or velocity of the fluid. The volume flow rate (usually called volumetric flow rate) is the quantity of fluid that passes by a point, say point A, in a set time, expressed for example in litres per second. Similarly, the mass flow rate is the mass or weight of fluid that passes by point A in a set time, such as kilograms per second. The velocity of the fluid is the distance travelled by the fluid (or the length of fluid that passes point A) in a set time, in metres per second.

If we have enough information, we can determine one of these parameters from another. For example, if we can measure the velocity of the fluid in the pipe and know the radius of the pipe r, we can calculate the volumetric flow rate. In the case of Figure 7.1 the volumetric flow rate will be the cross-sectional area of the pipe (which is $\pi \times r^2$) multiplied by the velocity of the fluid. If we know the volumetric flow rate and the density ρ of the fluid, then we can calculate the mass flow rate. The mass flow rate in Figure 7.1 is the volumetric flow rate multiplied by the density of the fluid.

Flow measurement is applied to liquids and gases, and occasionally with a collection of solids which have fluid characteristics when moving (for example, some types of sand, powder, or polystyrene packing chips). The range and diversity of applications of flow measurement is vast, from medical applications such as monitoring blood flow in the human body, recording gas use by means of domestic gas meters, to large-scale techniques in hydroelectric power stations.

It is important to choose the correct form of measurement to match the application. Volumetric flow rate may be used, for example, when measuring the amount of fluid filling a tank of a set size, say the amount of oil in an oil-tanker. However, mass flow rate may be more appropriate when filling an aircraft with fuel, because the range capability of an aircraft is often determined by the mass of the fuel, not the volume. Sometimes it may be the velocity of the flow that is important. Aircraft speed, for example, may be accurately estimated by measuring the pressure of the air flowing past the aircraft.

Volumetric flow rate

Although there are other methods available, there are four common techniques of measuring volumetric flow rate:

- Helical screw meter.
- The rotating lobe meter.
- The turbine meter.
- The paddle wheel meter.

These methods can usually be adapted to give an indication of velocity as well as flow quantity. Because they are in direct contact with the fluid, they disrupt the flow. Much of their accuracy depends on minimising this disruption, usually achieved by keeping friction low and making them light so they rotate freely.

The helical screw meter and rotating lobe meter are both 'positive displacement' meters, which means the fluid flows through chambers of known volume causing the screw or lobes to rotate. The principle of operation of positive displacement meters is to divide the fluid flow into known quantities (the size of the chamber) and then add these quantities together to find out the total quantity passed through in a given time. In practice, calibration information from the manufacturers will be supplied with the device, so that it can be set up to measure flow in the most accurate way.

Positive displacement meters have the following characteristics:

- They are accurate and can be used for either liquid or gas flow.

- They need little or no maintenance or recalibration.
- They are fairly expensive, can introduce significant pressure loss, and cannot usually measure quickly fluctuating flow rates.

Common applications of positive displacement meters are in petrol pumps, water meters and gas meters.

Helical screw meter

Helical screw meters are positive displacement meters which measure the flow of liquids highly accurately. A typical design of helical screw meter is shown in Figure 7.2.

The helical screw meter is mounted in a pipe, and the flow of the liquid through the meter causes the specially shaped rotor to turn. As the liquid flows through the meter, it is divided into known quantities by the rotor, because it is trapped in pockets formed by its helical shape.

A magnet is mounted on the rotor shaft with a pick-up coil placed just above it (an electromagnetic pick-up). As the magnet moves past the coil it induces a voltage pulse into the coil. Each induced voltage pulse is counted and, as the quantity of fluid which causes one revolution of the shaft is known, the total count allows the total quantity of fluid delivered to be determined.

Similarly, optical pick-up techniques can sense and detect the motion of the rotor. and produce a pulsed output. Light reflects off, or is interrupted by, the shaft (usually, when using the reflective technique, a special reflective strip is attached to the shaft to ensure good light reflection). The pulse rate of this reflected or interrupted light can then be used to measure the flow velocity and quantity.

Helical screw meters introduce significant pressure loss and are fairly expensive, but can be used with a wide variety of different liquids and flow rates. Some helical screw meters are bi-directional.

Rotating lobe meter

The rotating lobe meter is another type of positive displacement meter. Figure 7.3 shows a typical rotating lobe meter design. The lobes, positioned within the chamber at right-angles, will rotate synchronously when fluid flows, as shown in the diagram. Each lobe will trap a set amount of liquid during part of its rotation.

For each complete rotation of the lobes a known quantity of fluid passes through the chamber. At the end of each rotation a counter is incremented and, because the quantity that passes through the chamber is known, a measure of the total quantity delivered can be determined.

Accurate results can be achieved using a rotating lobe meter, and they may be used, for example, in metering fuel-oil deliveries to consumers.

If provided with power, the rotating lobe meter can also work as a pump which delivers fluid and provides a flow reading at the same time. This is an example of a positive displacement pump. These are used where accurate control of flow is needed, for example, a dosing pump in a chemical treatment plant.

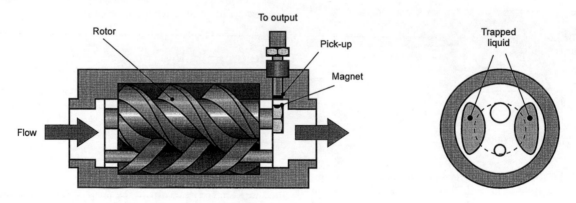

Figure 7.2 Helical screw meter

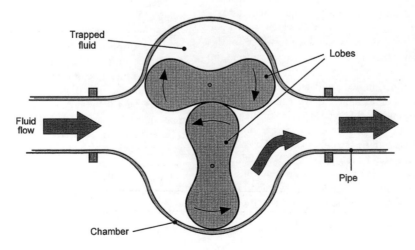

Figure 7.3 Rotating lobe meter

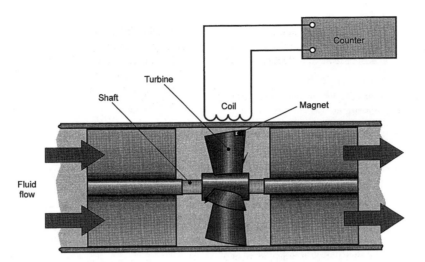

Figure 7.4 Turbine meter

Turbine flow meter specification

Line bore
28 mm (nominal)

Linear flow range
From 0.5×10^{-3} m^3.s^{-1} to 3.5×10^{-3} m^3.s^{-1} (from 0.5 to 3.5 litres.s^{-1})

Calibration factor K
Typically 150×10^3 pulses per cubic meter (150 pulses per litre)

Accuracy
Measured flow rate is within 0.5% of actual flow rate over specified linear range

Linearity
Not more than 0.5% from mean over linear flow range, based on liquid of density 10^3 kg.m^{-3}. 0.25% maximum deviation can be achieved over selected flow range

Repeatability
0.1% of actual flow rate over linear range

Over-range
To 150% of maximum rated flow for short periods

Minimum output voltage
30 mV peak-to-peak at minimum linear flow

Process fluid
The process fluid must be clean with light lubricating properties. *Note:* Clean water does not have the lubrication properties to permit operation of the meter

Response time
From 8 to 25 s for 65% response to step change

Pick-up coil
Body: stainless steel, sealed, threaded into flow meter housing
Coil: High-temperature insulated wire around permanent magnetic core
Impedance: 4500 at 600 pulses per second
Resistance: 3200 at 20% at 21 °C
Temperature limits: −200 and +200 °C

Pulse rate at maximum flow
Typically 550 pulses per second

Pressure drop on water at maximum flow
50 kPa

Figure 7.5 Turbine flow meter specification

Turbine meter

Turbine meters are used to show the volumetric flow rate and also the speed of fluid flow.

In a turbine meter, a typical example of which is shown in Figure 7.4, the fluid flow causes the turbine to rotate. The speed of rotation of the turbine is proportional to the velocity of the fluid. This may be determined by mounting a small permanent magnet at the tip of one or more of the turbine blades or on the turbine shaft, with a pick-up coil placed just outside the pipe. Similarly, optical pick-up techniques can sense each revolution of the turbine.

Turbine meters introduce some pressure loss and they are expensive, but also sensitive and highly accurate. They range in size from those designed to measure flows of fractions of litres per second, to very large meters measuring flows of hundreds of litres per second. Their accuracy and sensitivity depends on how easily the turbine rotates (the ease of rotation may be reduced by friction between the blades and fluid, or the shaft, or the bearings) and also on the nature of fluid and straightness of flow they are measuring. At very low flow rates the turbine meter may not respond correctly, and a minimum flow rate for accurate operation will usually be stated in the calibration information.

Figure 7.5 shows a typical turbine flow meter specification. Notice that it will produce 190 pulses per litre and can measure accurately up to 3.2 litres per second, but this particular model would not be suitable for water.

Paddle wheel meter

Paddle wheel meters, like the one shown in Figure 7.6, can accurately measure the flow of liquids.

The flowing liquid is formed into a jet, set so that it rotates the paddle wheel. A known quantity of liquid passes from the inlet to the outlet during each rotation of the paddle wheel, and so the flow rate can be determined. The paddle wheel may drive a mechanical counter (directly or by some form of linkage), or a pulsed output may be produced by magnetic or optical pick-up methods.

Paddle wheel meters introduce some pressure loss, are usually only suitable for liquids, but are less expensive than turbines. Some types of paddle wheel meter just indicate that flow is present and do not actually measure it.

Mass flow rate

A simple method of directly measuring the mass of a flowing liquid and so find out its mass flow rate is to use a gravimetric tank.

A gravimetric tank determines a known mass of liquid discharged into it over a certain time. The principle of operation of a gravimetric tank is shown in Figure 7.7.

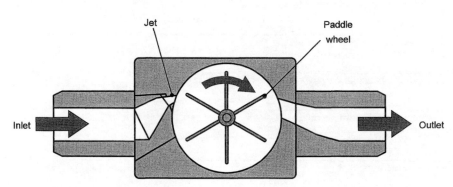

Figure 7.6 Paddle wheel meter

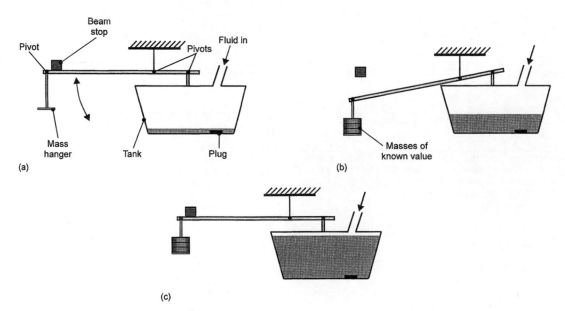

Figure 7.7 Gravimetric tank determining mass flow rate

In the design shown, the fluid is discharged into the tank. When the beam reaches the horizontal position, as shown in Figure 7.7(a) it hits a beam stop. At this point the fluid and the tank are equal in mass to the beam, and timing (using a stopwatch or automatic timing device) is started.

Fluid continues to discharge into the tank and masses of known value are attached to the mass hanger. The force produced by the masses and the acceleration due to gravity pulls the beam down as shown in Figure 7.7(b). Timing continues until the beam returns to the horizontal position; Figure 7.7(c). At this point the mass of the fluid in the tank is equal the force produced by the mass on the hanger, and so timing is stopped. The mass flow rate is then the mass of the fluid discharged into the tank (which is the same as the value of the known mass), divided by the time taken.

The beam will usually be pivoted off centre, a typical value for the ratio of the arm length is 3:1. This means that the mass of fluid in the tank will be three times greater than the masses on the end of the beam.

If the density of the fluid is known, the volumetric flow rate can also be determined. It is important to take temperature into account as this affects the density of the fluid.

Gravimetric tanks are not always as convenient to use as other flow rate sensors. However, this method of measuring the flow rate of a liquid is one of the most accurate techniques available. Often, other types of flow sensors are calibrated using a gravimetric tank. They are also used for laboratory and experimental work.

Other devices used to measure mass flow rate directly tend to be based around complex fluid mechanics and it is more usual to use a device which measures either volumetric flow rate or velocity, and convert these to mass flow rate. Some instruments which show mass flow rates measure the volume of flow and the density of the liquid or gas at the same time. They then calculate the mass flow rate from the measured volume and density, using a microprocessor.

Velocity measurement

The following flow meters measure the velocity of flow at a fixed point in a pipe or duct:

- Pitot-static tube.
- Hot-wire anemometer.
- Variable area flow meter.

Pitot-static tube

The Pitot-static tube, named after its inventor Henri Pitot, consists of a tube inserted into a fluid flow stream. It uses pressure tappings which show total pressure and static pressure to measure the fluid velocity. Static pressure is the pressure in a moving fluid which is unaffected by the fluid motion.

Figure 7.8 shows a design of Pitot-static tube used for measuring the velocity of a fluid flowing through a pipe. The total pressure tapping of a Pitot-static tube faces the flow and receives the total pressure from it. In Figure 7.8 the total pressure tapping is a tube positioned in the middle of the pipe, and bent so it faces directly upstream.

The static pressure tapping in is in the wall of the pipe, with its open end being perfectly smooth with the pipe wall so it causes no disturbance to the flow.

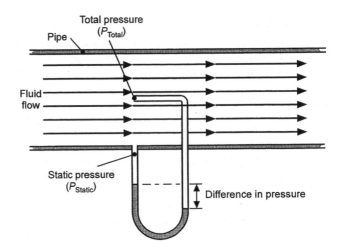

Figure 7.8 Pitot-static tube measuring the velocity of a fluid in a pipe

Figure 7.9 shows a Pitot-static tube used for measuring the velocity of air flowing through a duct. This works on exactly the same principle as the design of Pitot-static tube shown previously in Figure 7.8.

In Figure 7.9, which consists of two concentric tubes, the total pressure tapping of the Pitot-static tube is at the front of the inner tube. The static pressure tappings are holes drilled in the outer tube at right angles to the direction of the flow. Provided these holes are placed far enough from the tip of the probe, then the flow is undisturbed by the holes.

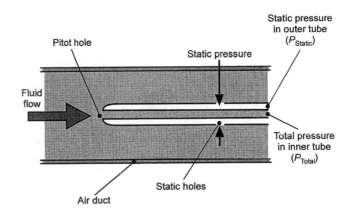

Figure 7.9 Pitot-static tube measuring the velocity of air through a duct

The pressure difference between the total pressure and static pressure tappings can be measured by a simple U-tube manometer, but the other techniques, such as piezometers, may also be applied. The difference between the total and static pressures is measured and the velocity, v, is given by

$$v = \sqrt{\frac{2\left(P_{\text{Total}} - P_{\text{Static}}\right)}{\rho}}$$

where:

- P_{Total} is the total pressure
- P_{Static} is the static pressure
- ρ is the fluid mass density

The Pitot-static tube can be used for both gas and liquid flow and provides good accuracy. It can measure low velocities as well as supersonic gas flows. One well-known application of the Pitot-static tube is in giving an indication of aeroplane air speed by applying the pressure difference to a Bourdon gauge, which is then calibrated in air speed units.

Hot-wire anemometer

Unlike the other devices we have discussed in this chapter which are used with both liquids and gases, the hot-wire anemometer is nearly always used for measuring the velocities of gases only.

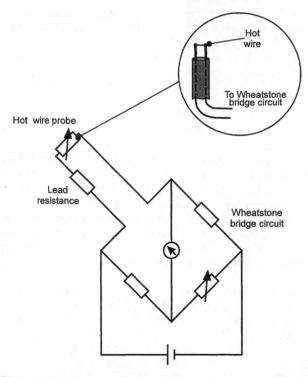

Figure 7.10 Hot-wire anemometer

An electrically heated wire probe forms part of a Wheatstone bridge circuit, as shown in Figure 7.10. The hot-wire anemometer works on the principle that gas flowing over a hot wire probe will cool it, and this rate of cooling is proportional to the gas velocity. On this basis, if a constant current is applied to the wire, because the resistance of the wire will change with temperature, the voltage will also change (because voltage is current multiplied by resistance). By measuring the voltage change in the Wheatstone bridge the velocity of the gas flowing over the wire can be calculated.

Alternatively, if the current is adjusted to keep the temperature of the wire constant, then similarly the value of the current will indicate gas velocity.

Hot-wire anemometers will have calibration information supplied with them, usually in the form of a graph, so the voltage or current reading can easily be related to gas speed. They can be used over a large range of flows, from very low (for example, 0.03 m.s^{-1}) to supersonic. They tend to be more expensive than Pitot-static tubes but can measure flows which are not as steady.

Variable area flow meter (rotameter)

A typical variable area flow meter is shown in Figure 7.11. It consists of a clear, tapered tube containing a float. The variable area flow meter mounts vertically in the pipeline carrying the fluid (note the general orientation of the pipeline does not have to be vertical, only the variable area flow meter).

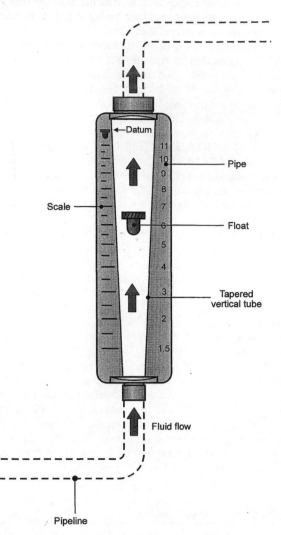

Figure 7.11 Variable area flow meter

Fluid flows through the gap between the float and the tube and this creates a pressure difference which forces the float up the tube. To keep the float in the centre of the tube, the float may have small vanes so it rotates, or have a fixed vertical guide rod running through the centre of it. At a steady rate of flow, the float remains stationary when the upward force is just enough to offset the weight of the float (that is, when the upward force and opposing weight of the float are in equilibrium). The height of the float in the tube is proportional to the fluid flow rate, which is read from a calibrated scale.

Variable area flow meters, (sometimes called rotameters) are used for measuring both liquid and gas flows. There are various designs of variable area flow meter and floats, depending on the type of fluid and the velocity range to be measured.

Variable area flow meters need little recalibration or maintenance, but their accuracy can be considerably affected by changes in temperature of the fluid they are measuring. Also, the skill of the reader in judging float position affects the accuracy of variable area flow meters, especially under circumstances where the float does not stay steady for long.

Constriction effect devices

Several types of device used to measure velocity of a fluid or volumetric flow rate are based on the 'constriction effect'.

Referring to Figure 7.12, a constriction in a pipe causes the fluid velocity to increase. Maximum velocity occurs at the minimum cross-section of the pipe.

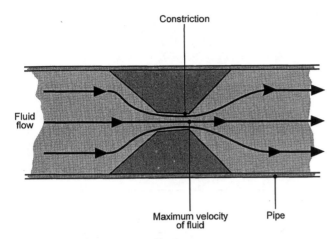

Figure 7.12 Constriction effect

As the velocity of the fluid in the pipe increases, the pressure at that point in the pipe decreases. Assuming the fluid cannot be compressed, the pressure difference between the pressure before constriction and the pressure at constriction depends on the following:

- Initial velocity of the fluid.
- Cross-sectional area of the pipe before the constriction.
- Cross-sectional area of the pipe at the constriction.

If we know the dimensions of the pipe both before and at the constriction, and if we can measure the pressure difference, we can then find out the velocity or volumetric rate of flow.

There are many forms of constriction flow meters. For flow in closed channels (such as pipes), the most commonly used, and the devices we shall consider here, are:

- The Venturi tube.
- The orifice plate.
- The nozzle.
- The Venturi nozzle.

Venturi tube

The Venturi tube is a device which has been used over many years for measuring the rate of flow along a pipe. It is named after Giovanni Battista Venturi (1746–1822), who performed experiments on flow in tapered tubes.

A typical Venturi tube is shown in Figure 7.13. It consists of a tapering contraction section, along which the fluid accelerates towards the constriction, followed by a section which diverges gently back to the original diameter (such a slowly diverging section is often referred to as a diffuser). As the velocity increases from the inlet section to the constriction, there is a fall in pressure. The magnitude of this fall in pressure depends on the rate of flow. The flow rate may then be calculated from the difference in pressure, as measured by manometers, piezometers, or other pressure gauge (see Chapter 3) placed at the constriction and upstream of it.

The flow velocity is directly proportional to the square root of the pressure difference and thus the indicator may be calibrated directly in velocity units, volumetric flow rate units or mass rate of flow units.

Remote reading can be achieved by using pressure sensors with electrical output. The two signals are subtracted and the square root of the result taken, before matching the result to a recorder or display (or perhaps producing a control signal). We shall see how this is done in Chapter 9.

A manufacturer's specification for a range of available Venturi tube flow meters is shown in Figure 7.14.

Compared with nozzles and orifice plates, Venturi tubes are expensive and need more space in a pipeline. However, they give a significantly smaller pressure loss and can be used on types of fluid which nozzles and orifice plates cannot be.

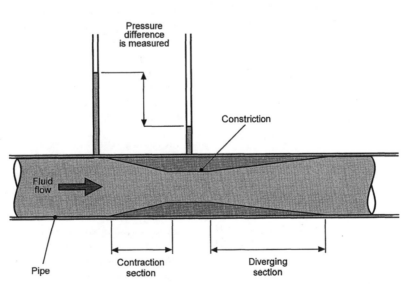

Figure 7.13 Venturi tube

Venturi tube flow meters

Materials

Venturi tube flow meters are available in a range of materials, including: aluminium, brass/bronze and AISI 316 stainless steel

Accuracy

In accordance with British Standard BS 1042. (Typically ±1% of full scale. Calibration on liquids and gases to better than ±0.2% of actual mass flow.)

Flow range

Water at 15 °C: any 10 : 1 range between 0.83×10^{-3} $m^3.s^{-1}$ (50 litres.min^{-1}) and 41.7×10^{-3} $m^3.s^{-1}$ (2500 litres.min^{-1}). Air at STP: any 10 : 1 range between 10^{-2} $m^3.s^{-1}$ (600 litres.min^{-1}) and 0.5 $m^3.s^{-1}$ (30000 litres.min^{-1})

Nominal bore

65 mm

Maximum recommended measured pressure difference at maximum selected flow

Liquids: 147 kPa (15 meters water gauge)
Gases: 14.7 kPa (1.5 meters water gauge)

Overall pressure drop (head loss)

Usually designed to be one tenth of maximum measured pressure difference

Maximum kinematic viscosity

10^{-2} $m^2.s^{-1}$

Maximum temperature/pressure rating

Aluminium: 5 MPa (50 bar) up to 150 °C
Brass/Bronze: 10 MPa (100 bar) up to 150 °C
AISI 316 Stainless Steel: 7 MPa (70 bar) up to 600 °C

General data required

Type of fluid, flow range, static pressure, maximum differential pressure required, working temperature, specific gravity, operating density for gas flow, percentage moisture for steam flows, operating viscosity for liquid flows

Figure 7.14 Typical specification for Venturi tube flow meters

Orifice plate flow meter

An orifice meter, as shown in Figure 7.15, typically consists of a disc with a hole in the centre which is inserted into the pipe. The fluid emerges from the orifice plate as a convergent stream which is projected into the pipe. The pressure difference is measured between a point a distance d, upstream from the disc and a point $d/2$ downstream, where d is the diameter of the pipe. The $d/2$ downstream distance is required because this is where the highest pressure occurs. This point is called the 'vena contracta', and the streamlines are parallel at this point.

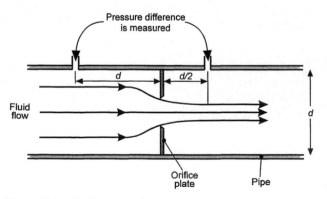

Figure 7.15 Orifice plate flow meter

The flow rate may then be calculated from the difference in pressure between the two points.

Direct or remote measurements can be taken from the orifice plate, in the same way as the Venturi tube. Both may be used for gas or liquid flow, but the orifice plate method results in a much greater pressure loss largely because of eddying created by the plate. It also needs a longer straight section of pipe upstream of it than required by the Venturi tube.

Of the Venturi tube, orifice plate and nozzle, the orifice plate is the most commonly used because it is a low cost device, simply constructed, and easily inserted into existing pipelines.

Nozzle

The nozzle, as shown in Figure 7.16, is similar to the orifice plate in that a convergent stream of fluid emerges from it and is projected into the downstream pipe. Note the pressure tappings are immediately before the nozzle, and directly after where it joins the pipe. The flow rate is calculated from the difference in pressure between the two pressure tappings, and calibration is usually in volumetric flow rate.

Nozzles tend to need less straight pipeline before them than orifice plates, and they are short and need less installation space than Venturi tubes. The pressure loss caused by a nozzle is usually less than an orifice plate but still significant when compared with a Venturi tube.

Nozzles can be used for liquid and gas flows and provide good accuracy. Direct or remote measurement can be taken from the nozzle, in the same way as the Venturi tube and orifice plate.

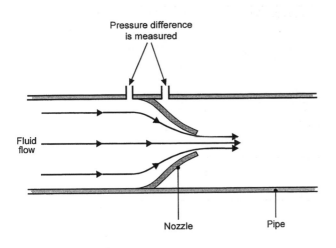

Figure 7.16 Nozzle

Venturi nozzle

To reduce the pressure loss caused by the nozzle, while keeping some of the convenience and simplicity, there is another type of nozzle called the Venturi nozzle. A Venturi nozzle design is shown in Figure 7.17. It is basically a cross between the standard nozzle and the Venturi tube, in that the upstream end consists of a nozzle inlet. However, this is followed by a section which diverges gently back to the downstream pipe diameter, like the Venturi tube.

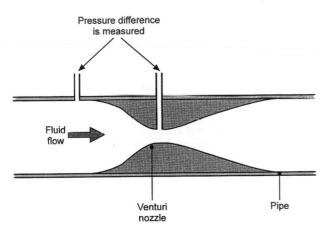

Figure 7.17 Venturi nozzle

Like nozzles and Venturi tubes, Venturi nozzles can be used for liquid and gas flows and provide good accuracy. Direct or remote measurement can be taken from the Venturi nozzle, in the same way as the Venturi tube, orifice plate, and nozzle. Venturi nozzles are usually more expensive than standard nozzles, but cause significantly less pressure loss. They are easier to install and less expensive than Venturi tubes.

Non-disruptive flow measuring devices

The flow measuring devices we have considered so far all disrupt the flow they are measuring to some extent. In some cases it may be undesirable, or not possible, to disrupt the flow to be measured, for example in some medical applications. Hence devices which do not come into physical contact with the fluid and therefore cause no disruption have been developed.

Non-disruptive flow measuring devices tend to use electronic techniques to sense the flow, are generally more accurate than disruptive techniques, but are usually more complex and significantly more expensive.

There are several types of non-disruptive flow measuring devices available. Here we'll look at two basic methods:

- The electromagnetic flow meter.
- The ultrasonic flow meter.

Electromagnetic flow meter

The electromagnetic flow meter is based on the application of Faraday's Law of Electromagnetic Induction.

To apply Faraday's Law of Electromagnetic Induction to measure the flow rate of a fluid, the electromagnetic flow meter uses the fluid as the conductor. By measuring the size of an induced e.m.f. the flow rate can be calculated. The size of the e.m.f. may be determined from the expression,

$$e = Blv$$

where:

- e is the induced e.m.f. (measured in volts)
- B is the magnetic flux density (measured in teslas)
- l is the length of the conductor cutting flux (measured in metres)
- v is the velocity of the conductor (measured in $m.s^{-1}$)

A typical electromagnetic flow meter is shown in Figure 7.18. Two coils, connected in series, are mounted either side of the pipe at right angles to the direction of flow. When the coils are energised by a current a small magnetic field is created across the fluid.

Two conductance probes are mounted to be in direct electrical contact with the fluid (but insulated from all other components). They are positioned at right angles to the fluid flow and the magnetic field. These probes sense the e.m.f. induced and this signal is conditioned so it can be used by external displays and recording instruments.

Depending upon the specific application, the output may be calibrated in terms of the mean velocity of the flow, the volumetric flow rate or, for fluids of constant density, the mass flow rate.

Because of the weak magnetic field the magnitude of the induced e.m.f. is extremely small, and this may lead to problems with electrical noise and significant errors in the measurement. However, they do not disrupt the flow and are not affected by fluids containing suspended matter, such as liquid cement or food pulp. Other applications of the electromagnetic flow meter vary from the flow measurement of liquid metals to blood flow rate monitoring.

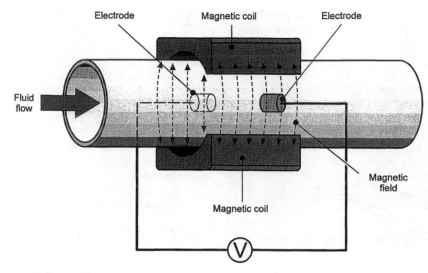

Figure 7.18 Electromagnetic flow meter

Electromagnetic flow meters may also be used to measure the speed of a ship moving through water. Here, the flow is on the outside of the sensor and not through it. A submersed electromagnetic probe assembly is mounted beneath the hull of a ship. Coils carrying current create a magnetic field in the water surrounding the assembly so that as the ship moves the conductive water passes through the field. With the probes positioned at right angles to both the field and the direction of flow, as before, the induced e.m.f. may be sensed and used to give an indication of the speed of the vessel. The speed measurement is continuous and linear though some adjustment is required to allow for variations in the salinity and temperature of the water. Electromagnetic flow meters cannot be used with gases.

Ultrasonic flow meter

Figure 7.19 shows a typical ultrasonic flow meter clamped onto a pipe. Here, the ultrasonic flow meter comprises an ultrasonic transmitter, a reflector and a separate receiver mounted a set distance away along the pipe.

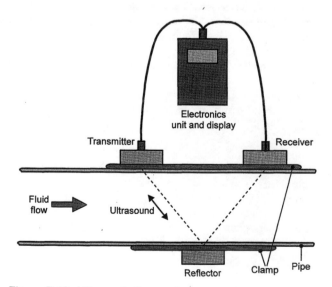

Figure 7.19 Ultrasonic flow meter

The transmitter produces a sound pulse, which rebounds off the reflector and finally reaches the receiver. If the flow is zero, the time it takes from the pulse being transmitted to reaching the receiver is controlled only by the distance between the transmitter and the receiver and the speed of sound in the fluid. However, if the fluid is flowing from the transmitter towards the receiver then the sound pulse will travel faster. The time delay between transmit and receive will then be smaller. Conversely, if the flow is in the opposite direction then the delay is longer. The electronics unit and display is calibrated so that the time delay is shown in terms of the flow rate of the fluid.

An alternative ultrasonic technique of measuring flow rate is based on the Doppler effect. The Doppler effect is commonly illustrated by the sound of a train or a vehicle approaching an observer. When the train is approaching the apparent sound pitch is higher than the actual pitch and when it is moving away the apparent sound pitch is lower. The difference between the actual and the apparent pitch is an indication of the speed of the vehicle.

In the case of the Doppler effect ultrasonic flow meter, increased flow rate from the transmitter to the receiver increases the apparent pitch of the received signal in direct proportion. Measurement of this change in frequency may be used to indicate the flow rate.

Although expensive, ultrasonic flow meters are highly accurate and stable. They can be used with many liquids, conducting and non-conducting and can measure flow continuously in both directions They do not disrupt the flow, and are often portable. However, particles of suspended matter in the liquid are required for either ultrasonic technique to work successfully. Ultrasonic flow meters cannot be used with gases.

Summary

In this chapter we have looked at three ways in which flow is quantified, that is, volumetric flow rate, mass flow rate, and velocity, and looked at some of the devices and techniques commonly used to measure flow. All of the devices we have looked at are available in varying sizes and calibration ranges, the choice of which depends on the measurements required by their intended application and the limitations of the device.

There are other devices, techniques, and adaptations of the ones we have discussed, in use. There are also other, less common, methods of quantifying flow. This chapter will have provided you with a good insight into the subject of flow measurement.

Questions for further discussion

1. Look in detail at the turbine flow meter specification shown in Figure 7.5. What practical application or applications would the device specified be particularly suited for?
2. Consider the Pitot-static tube shown in Figure 7.9, which is designed for use in air. If this design was to be used for a liquid, what would be the drawbacks and under what conditions would the best results be achieved?
3. What is the main disadvantage of using the constant wire temperature method to measure gas velocity in a hot-wire anemometer?
4. What special consideration needs to be applied to gases and not to liquids when measuring the pressure difference in constriction devices and hence calculating flow rates?

---

8 Display and Recording

The final element in the measurement system is the device used for displaying the measured values, or recording them to become a permanent copy for analysis at a later date. Many locally read sensors display measured values directly, because their display is an integral part of the sensor and they need little if any extra display equipment. Examples of these sensors are liquid-in-glass thermometers, spring balances, manometers, and dial test indicators. Most electronic devices need some form of additional display or recorder, such as a voltmeter, ammeter, oscilloscope, chart recorder or data logger. As we have seen, nearly all measurands can be represented as an electrical signal by choosing a suitable transducer and appropriate signal conditioning.

Like many systems, the overall quality of output from a measuring system is defined by its lowest quality element. If judged by the error of the displayed value, it is the product of the error of all of the component parts of the system. For example, we may use a highly accurate, precise and sensitive LVDT to measure displacement, yet display its output on a voltmeter whose accuracy, precision and sensitivity are significantly less. Here, the overall output of the system will only be as accurate, precise, and sensitive as the voltmeter specification, including the smaller error of the LVDT. Therefore it is important to select a display or recording device which is compatible with the rest of the system, giving the quality of output appropriate to the requirements of the user.

The output from most electric devices will be in the form of an electrical signal. A signal is a general term which refers to something which conveys information, whether it be a series of electrical pulses to a computer, a radio signal, or an electrical voltage.

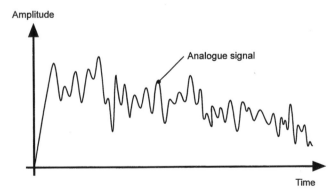

Figure 8.1 Analogue signal

There are two common types of signal that you will meet in measurement systems. These are the analogue signal and the digital signal.

Electrical analogue signals vary continuously in amplitude and time, as illustrated in Figure 8.1.

Electrical digital signals in measurement systems are usually in binary, which means each pulse will be at one of two voltage levels. Figure 8.2 shows a binary digital signal, whose amplitude corresponds to either logic level 0 (low) or logic level 1 (high).

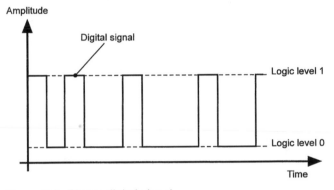

Figure 8.2 Binary digital signal

There are many display and recording devices available, in a wide range of cost, size, and accuracy. They can be conveniently split into the following general categories:

- Analogue displays.
- Digital displays.
- Recorder units.
- Computers and data loggers.

Analogue displays

A display is a device which gives an instantaneous visible indication of the signal from the sensor, but does not save it for future use or reference.

Displays are real-time devices. The term real-time means something which is happening in the present. For example, if you are watching a live television broadcast, you could say you are watching it in real-time, even though there is a small transmission delay between actual events and the signal being displayed on your television set. A real-time device is a device which responds with minimal delay to the events which influence it.

Analogue displays rely more on the skill and interpretation of the reader than digital displays. However, it is generally easier to determine trends using an analogue rather than a digital display, and analogue displays sometimes require less signal conditioning.

Moving coil meter

The moving coil meter is a display device used as the basis of analogue voltmeters, ammeters, multimeters, and certain specialist applications. Figure 8.3 shows the front panel of a typical moving coil voltmeter.

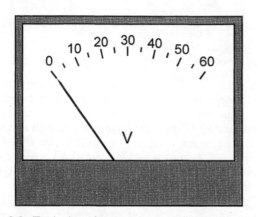

Figure 8.3 Typical analogue voltmeter display

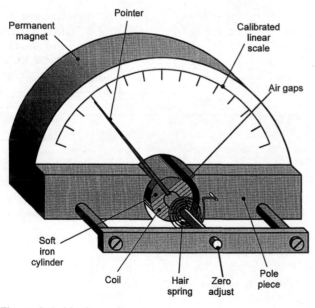

Figure 8.4 Moving coil meter

The basic construction of a moving coil meter is shown in Figure 8.4. It consists of a soft iron cylinder with a coil suspended around it. The coil is pivoted between two magnetic pole pieces and connects to a pointer mechanism. The coil can move freely in the air gap between the pole pieces and the soft iron core. The pole pieces are shaped to produce a uniform magnetic field in the air gap. When current flows in the coil it reacts with the magnetic flux to cause the coil to rotate. The pointer movement is proportional to the magnitude of the voltage or current being measured. A hair spring opposes the movement of the pointer.

When the torque produced by the spring is equal to the torque produced by the coil, the pointer settles on the scale at a value appropriate to the current or voltage.

Moving coil meters connect to the sensor output by means of test probes, or may be built in to a circuit and mounted on a panel.

The electrical rating of the resistance of the coil is one of the main considerations when designing a moving coil meter to measure either current or voltage. Generally if they are designed to measure voltage the coil will have a high resistance, and if measuring current a low resistance.

Moving coil meters are inexpensive and reliable, and are available in a wide variety of ranges. The meters themselves can be made to a high degree of accuracy, but their overall accuracy is limited by the skill of the reader. Response is relatively slow and they can usually only measure currents or voltages which are d.c. though they can be bi-polar.

Resistance meter

The resistance meter, also known as the ohmmeter, is as its name suggests an instrument used to measure electrical resistance. It is based on the moving coil meter and Figure 8.5 shows the front panel of a resistance meter.

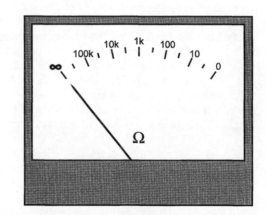

Figure 8.5 Resistance meter

The basic resistance meter consists of a moving coil meter with a built in power supply. The resistance to be measured is connected by means of test probes.

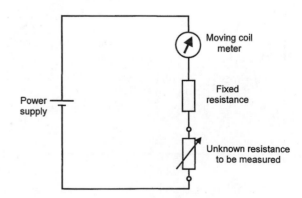

Figure 8.6 Resistance meter circuit

Figure 8.6 shows the basic electrical resistance meter circuit. The resistance to be measured is placed in series with the moving coil meter, so that it completes the circuit. The current through the meter is inversely proportional to the resistance being measured, and the meter can be calibrated in terms of resistance (ohms). The scale is a reciprocal scale which is non-linear. Most digital resistance meters are also based on this circuit.

Connecting the test probes of the resistance meter together will cause full scale deflection of the pointer, which corresponds to zero resistance. When the test probes are not connected to anything, the meter will read infinity (∞). Hence resistance meter scales are in the reverse direction to current or voltage scales.

Resistance meters are accurate and particularly suitable for measuring small resistances. Because of the non-linear scales, at the higher end of the scale they are more difficult to read and less accurate. The accuracy also depends on the stability of the power supply, and frequent recalibration is often required.

Moving iron meter

Externally, moving iron meters may be similar in appearance to moving coil meters. However, their principle of operation relies on the fact that when magnetised due to a current, soft iron can be made to attract or repel. As the current increases, the magnetic flux of soft iron becomes proportionally stronger.

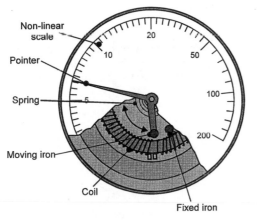

Figure 8.7 Moving iron meter

Figure 8.7 shows the principle of a moving iron meter. It consists of a fixed coil and two pieces of soft iron which become magnetised when current flows in the coil. One piece of iron is fixed, and the other piece is joined to a pivoted pointer mechanism. When the pieces of iron become magnetised, they repel, causing the pointer to move. Like the moving coil meter, a hair spring opposes the movement of the pointer. When the torque produced by the spring is equal to the repelling force produced by the magnets, the pointer stops on the scale at a value appropriate to the current or voltage.

Moving iron meters measure a.c. or d.c. voltages and currents. They lend themselves to applications where large, non-linear scales are appropriate, for example where the signal may surge suddenly

Moving iron meters are generally less expensive than moving coil meters, but are not as accurate.

Cathode ray oscilloscope

The cathode ray oscilloscope, as well as many other applications, is used for detailed monitoring of electrical output signals from transducers. Usually referred to simply as the oscilloscope, it displays electrical signals as waveforms with high accuracy and precision. Figure 8.8 shows a typical oscilloscope, and Figure 8.9 shows a typical waveform as it may appear on an oscilloscope display. Note that time is on the x-axis (horizontal) and the parameter being monitored, in this case voltage, is on the y-axis (vertical). The signal shown in Figure 8.9 appears as a sine wave because it is displayed against time on the x-axis. If we did not use this function of the oscilloscope, the waveform would only vary in amplitude.

The cathode ray oscilloscope displays the waveforms on a cathode ray tube. The cathode ray tube is also found in television sets and computer monitors. It consists of an evacuated tube along which passes a beam of electrons as shown in Figure 8.10. The beam of electrons is produced from an electron gun and strikes a fluorescent screen to produce a small spot of light on the screen. The beam can be made to deflect in both the x and y directions, by potential differences applied to two sets of plates between which the beam passes.

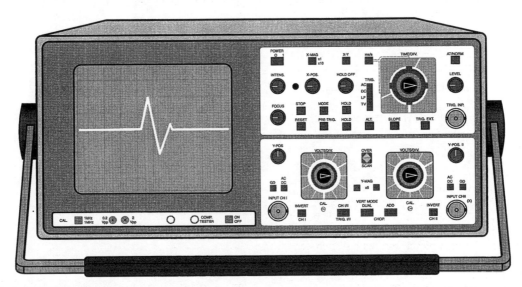

Figure 8.8 Cathode ray oscilloscope

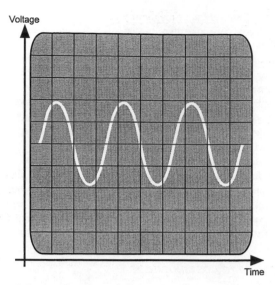

Figure 8.9 Oscilloscope display

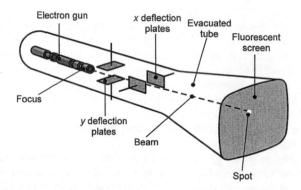

Figure 8.10 Cathode ray tube

The output signal from the transducer is usually accessed via probes or clamps connected to the oscilloscope. The signal from the transducer is fed to the *y* plates and a steadily increasing voltage signal connects to the *x* plates. This signal, called a linear ramp, deflects the beam across the tube at a fixed rate and thus provides an axis which can be calibrated with respect to time. The speed that the spot moves across the screen can be high enough so it appears as a continuous line of light.

> **Key fact**
>
> A ramp signal is a voltage which rises steadily, as in a saw-tooth waveform.

The oscilloscope has controls to adjust the sensitivity of both the *y* and *x* scales, for control of brightness and focus, and for synchronising the *x* and *y* deflections. The front of the oscilloscope screen usually has a scaled grid (a graticule) over it, allowing easy reference to the *x* and *y* co-ordinates. By careful selection of circuits and scales, the waveform can be displayed on the screen to be an accurate representation of the signal against time. From this display the shape of the wave, frequency, amplitude and other information can be obtained.

There are many models of oscilloscope available. For example, dual trace oscilloscopes have two beams for comparison of different waveforms, and storage oscilloscopes can freeze a transient waveform.

Oscilloscopes are widely used for testing equipment, and in research and development. They are relatively expensive compared to other display devices, and the high level of accuracy and precision they provide may not be necessary in many applications. However, their ability to give a detailed visual representation of an electrical signal makes the oscilloscope a very useful piece of display equipment.

Digital displays

Rather than showing an output value from a transducer in terms of meter deflection or a waveform, digital displays show the values as a number. The numerical value of the display is directly proportional to the size of the quantity being measured. Figure 8.11 shows a typical digital multimeter, using a digital numerical display.

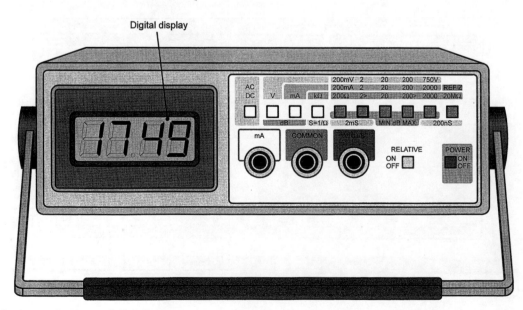

Figure 8.11 Instrument using a digital display

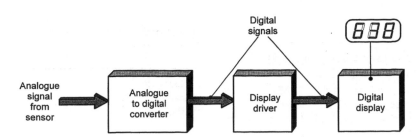

Figure 8.12 Flow diagram of a digital display

Digital displays are used in similar applications to analogue displays, for example digital ammeters, voltmeters, and multimeters. Operating these devices is similar to operating their analogue counterparts, and semiconductor technology allows the manufacture of small, light and convenient meters. Some meters show signals using a combination of analogue and digital displays, for example hand-held oscilloscopes.

All digital meters have a refresh rate. A refresh rate refers to how often the data displayed is updated. This does not apply to analogue signals which are updated continuously and effectively have an infinite refresh rate. It is an important factor affecting the quality of digital displays.

The electronic circuitry which converts the output signal from a transducer into a visual digital representation uses an analogue to digital converter and display driver, as shown in Figure 8.12 (analogue to digital converters are discussed in more detail at the end of this chapter). In this section we'll look at the ways in which different types of digital displays are presented.

A common way in which digital displays form characters is by using the seven-segment display.

Key fact

A seven-segment display is a method of forming the characters 0 to 9, by selectively highlighting and blanking appropriate segments arranged in the form of an eight.

A seven-segment display, showing which segments are highlighted and which segments are blanked off to form the numbers 0 to 9, is shown in Figure 8.13.

The seven-segment display is limited in that, although it can show the numerals 0 to 9, it has a limited range of alphabetic characters. The alphabetic characters which can be produced by the seven-segment display are not always a good representation and are a mixture of upper and lower case. To overcome this, the sixteen-segment display can be used, which is by no means perfect but can display all alphabetic as well as numeric characters. Figure 8.14 shows the basis of the sixteen segment display.

Light emitting diode

A light emitting diode (LED) is a diode which glows when supplied with a specified current. They are used in many applications, for example the optical sensors we met in Chapter 3, on or off indicators on computers, and safety lights on bicycles. They are limited in maximum size compared with normal incandescent lamps (the standard maximum LED size is about 10 mm diameter), but they can be made very small.

Light emitting diodes are normally made from compound semiconductors such as gallium phosphide or gallium arsenide, to which are added small controlled amounts of impurities. The colour of light emission depends on the diode material and the impurities used. Green, red, orange and yellow are common LED colours. Blue LEDs are available but are significantly more expensive than the other colours.

Single LEDs are often used in digital displays to indicate on or off, or high or low signal levels. For displaying more complex information, such as numbers, they are used in multiples.

In an LED display, the light emitting diodes are arranged in an array. By selective illumination, simple characters can be formed on the display. The principle of this is shown in Figure 8.15.

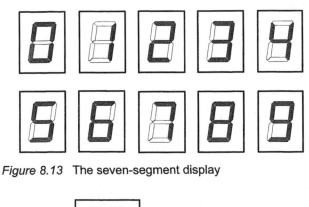

Figure 8.13 The seven-segment display

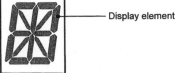

Figure 8.14 Sixteen segment display

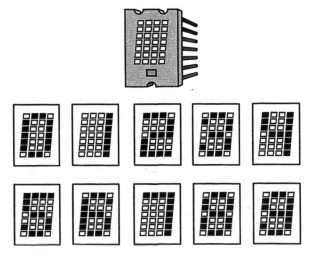

Figure 8.15 The LED display

To produce alphanumeric or more complex characters they can be arranged in a more complex array. Alternatively, if the display is kept fairly small the LEDs themselves can be shaped in the form of segments of a seven- or sixteen-segment display.

Light emitting diode displays are small, inexpensive and reliable. They have a fast response time, and can be made fairly bright, so are used in applications where the ambient light is low. They are often used in indicators and displays on control panels, for example when monitoring the speed of a generator. High numbers of LEDs arrayed in groups can produce a large display, for functions such as displaying temperature in a street or large factory.

As liquid crystal technology advances, the use of LED displays has become less. In many applications they have been replaced by the liquid crystal display which has a much lower power consumption. They are still used where initial cost is a more important consideration than power consumption.

Liquid crystal display

Liquid crystal displays are commonly referred to in their abbreviated form as LCDs. Figure 8.16 shows a seven-segment liquid crystal display. Each segment consists of a film of liquid crystal sandwiched between two transparent electrodes. Applying a potential difference across the electrodes causes the refractive index of the liquid to change and the segment appears opaque. Selectively energising individual segments forms different characters. Alphanumeric or more complex characters are produced by using a sixteen-segment display or a dot-matrix array.

Sometimes LCDs have the segments and background opaque and form numbers by making selected segments transparent.

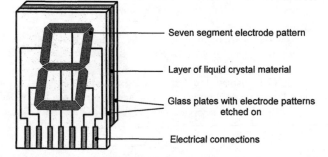

Figure 8.16 Seven-segment liquid crystal display

The main advantage of LCDs over other types of display is their relatively low cost and low power consumption. This is a major benefit if a display device is powered by batteries. However, their response time is much slower than the LED display. A typical response time for an LED display is 100 nanoseconds, compared with 10 milliseconds for a liquid crystal display. Also, liquid crystal displays need a source of illumination, especially if used where the ambient light conditions are low. They can be backlit but this increases overall power consumption.

Liquid crystal displays are nowadays the most widely used form of digital display. Applications include digital wrist-watches, voltage and current meters, and personal organiser display screens.

Mechanical digital displays

Mechanical digital displays are used where large displays are needed. They usually consist of a seven- or sixteen-segment display. Figure 8.17 shows a mechanical seven-segment digital display.

Bars shaped to fill each segment have one face which is the same colour as the background, and one brightly coloured face. The background colour is usually dark or black to contrast with the bright colour on the bar. The bars can rotate, so that each segment is either the same colour as the background, and so cannot be seen, or filled with a bright colour. Thus by selectively rotating the bars so some segments are coloured and others are blank, characters can be produced on the display.

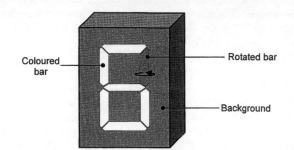

Figure 8.17 Mechanical seven-segment display

The bars are rotated by an electrical device such as a solenoid. A solenoid is a device which in this case causes a mechanical rotation when a current is applied to it.

The bars may be triangular or square in shape so that they have more than two useful faces. This allows different colours to be used.

Mechanical digital displays need a greater power source than LCD or LED displays, although they generally only consume power while the segments are changing. However, they can be made very large and so be read at long distances. Typical applications are public clocks, or information display panels at railway stations and airports. Because they include moving parts, they are not as reliable as other displays, requiring regular maintenance. Their response time is also slow, taking up to half a second to change state.

Numerical indicator tube

The use of numerical indicator tubes was once fairly common but is nowadays very rare. Their principle of operation is discussed here because you may occasionally come across one in old display equipment.

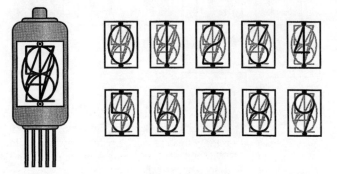

Figure 8.18 Numerical indicator tube

A basic numerical indicator tube is shown in Figure 8.18. The display is a gas-filled tube containing an anode (positive terminal) in the form of a mesh and ten cathodes (negative terminals) shaped in the form of the numbers 0 to 9.

A large potential difference across the anode and one of the cathodes causes that cathode to glow. Hence applying a potential difference across an appropriate cathode displays a number.

Unlike seven-segment displays, the numbers in a numerical indicator tube can be formed into a true shape. The display is much brighter than LED or LCDs, and can be read in full daylight or under fairly bright internal lighting conditions. However, they cannot be seen clearly at angles of more than about 45° from the front, and they are relatively expensive. They also need a much higher supply voltage than LED or LCDs (about 180 V) to make the cathode glow. The other display technology we have discussed has largely superseded the numerical indicator tube, and its use is restricted to maintenance of old equipment.

Recorders

A recorder is a device used to make a permanent record of a measured value or signal. This can be in the form of a graphical or numerical printout (hard copy), or as an analogue or digital signal recorded on to tape, disk, or semiconductor memory. These copies are useful for analysing and monitoring data, for example for monitoring wear in a manufacturing machine over a period, and so predict when a tool needs changing. Some parameters change so fast a real-time display is inadequate and hard-copy of data is essential. Other processes may be so slow it would be difficult or impractical for a person to monitor events in real-time, for example if recording environmental trends such as air temperature.

Although the advances in computers have reduced the requirement for some types of recorder, recorders will probably remain in use for the foreseeable future in various applications. A selection of common types are discussed here.

Moving coil recorders

Moving coil (sometimes called galvanometric) chart recorders are a simple and low-cost method of recording data in graphical form. They are based on the same principle of operation as the moving coil meters we looked at earlier in this chapter.

Figure 8.19 shows the basis of a moving coil recorder. The mechanism consists of a coil wound around a pivoted former. The coil former connects directly to a pen or marker. The electrical signal from the transducer whose output is being measured flows through the moving coil. When an electrical signal flows through the coil, the coil, former and hence the pen or marker deflect in proportion to the size of the current.

A chart driven by a synchronous motor moves at a constant speed. The pen or marker is in direct contact with the chart paper, and so draws a trace on it. This produces a graph of the signal current variation against time.

The zero position for the pointer is usually the centre of the chart, allowing recording of both positive and negative movements.

One of the main problems with this recorder is that the pen moves in an arc, and so does not traverse the paper in a straight line. The error caused by this arc can be reduced by using a long pen arm, or by having a scaled grid on the paper drawn in the same arc as the one swept by the pen. The signal causing the moving coil to deflect must be strong enough to overcome any

friction between the pen and the paper. Ink based pens such as felt tips have a relatively high degree of friction, so heat sensitive or electrosensitive markers and paper may be used.

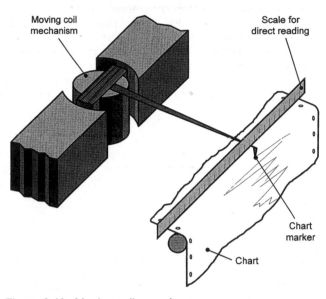

Figure 8.19 Moving coil recorder

Moving coil recorders have a relatively slow response time and are not suitable for high speed recording. This makes them suitable for recording data such as temperature or air pressure.

Servo chart recorders

Servo chart recorders are a common, and accurate way of recording data. Figure 8.20 shows a typical servo chart recorder.

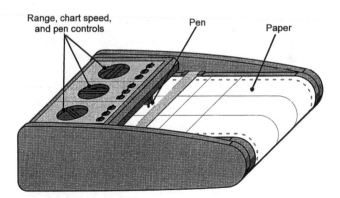

Figure 8.20 Servo chart recorder

Servo chart recorders use a pen, often a fine felt-tip, to draw a trace proportional to the signal being monitored. The pen, driven by a belt connected to a servo motor, traverses the paper linearly along a guide rail. The pen connects to the wiper of a linear potentiometer (which we met in Chapter 3) which monitors the pens displacement. An electrical circuit compares the output from the linear potentiometer to the signal from the transducer. The difference in voltage is the error signal. The circuit then drives the servo motor, which moves the pen, until the error signal is zero. Hence a trace proportional to the input signal is produced. This is an example of a closed-loop system, as discussed in Section 2.

Servo chart recorders are versatile and easy to use. Any friction from the pen is overcome by the electronic circuitry. They are small, portable, and can accommodate multichannels and use different coloured pens for clear traces. Their overall accuracy is limited by the accuracy of the potentiometer, but mainly by the pen. The pens, if not checked and changed regularly, can run dry of ink and so produce a faint or no trace, or occasionally put excessive ink on the paper.

A typical specification for a dual channel servo chart recorder is shown in Figure 8.21.

Dual pen chart recorder technical specification
Range
1 mV to 500 V d.c.
Pen speed
400 mm.s^{-1}
Chart speed
10 mm.hr^{-1} to 600 mm.hr^{-1}
Response time
0.5 s
Accuracy
0.5%
Size
140 × 330 × 335 mm
Power requirement
240 V a.c. 50–60 Hz

Figure 8.21 Typical specification for a dual channel servo chart recorder

XY plotter

The XY plotter, also known as the flat bed plotter, is an analogue device which produces a graph showing the relationship between two input signals. Figure 8.22 shows an XY plotter.

On an XY plotter, the paper is held in a fixed position on a flat bed. One signal voltage is applied to the *x*-input terminal and the other is applied to the *y*-input terminal.

The pen, which is usually a fine felt tip, is held on a carriage which traverses a bar. The bar can move across the paper, bottom to top. The voltage on the *x*-input terminal displaces the carriage, and hence the pen a proportionate distance left to right. The voltage on the *y*-input terminal displaces the bar, and hence the pen a proportionate distance bottom to top. Servo motors power both the pen carriage and the bar.

A typical application of the XY plotter is, for example, in plotting the stress-strain curve during a tensile test on a mechanical specimen Here, strain is input to one channel and stress the other.

The XY plotter can also plot variations of transducer signals with time. To do this, the *x*-input is in the form of a voltage ramp with a constant gradient (like the cathode ray oscilloscope). The signal to be recorded is connected to the *y*-input. The gradient of the ramp is chosen to suit the rate of time variation of the signal and the higher the signal frequency, the steeper the ramp.

XY plotters have the advantage over moving paper recorders in that they do not necessarily have to record a variable against time. However, they are fairly slow, and correct positioning of the paper, maintenance and quality of pens used is essential to achieve accuracy. If connected to a computer, graphs can be neatly labelled because the pen can be used to plot alphabetical and numerical characters. They can also plot drawings and graphs directly from computers. However, they have mostly been superseded by higher quality laser printers.

Ultraviolet light recorder

Figure 8.23 shows the basis of an ultraviolet recorder. The ultraviolet light recorder detects the signal from the transducer using a moving coil, in the same way as the moving coil meter and the moving coil recorder. However, instead of a pointer, the ultraviolet recorder uses a tiny mirror to reflect a beam of ultraviolet light emitted by a mercury-vapour lamp. This beam is reflected onto ultraviolet sensitive paper and focused by a mirror (or mirrors). This produces a trace of the input signal (current or voltage) amplitude against time.

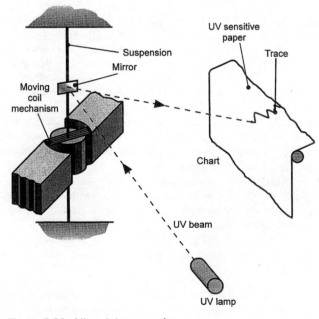

Figure 8.23 Ultraviolet recorder

Figure 8.22 XY plotter

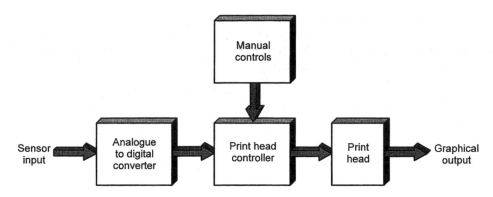

Figure 8.24 Thermal array recorder

Because the ultraviolet recorder mechanism is small and light, it needs very little force to move it. Hence it is sensitive in comparison to, say, the moving coil meter and can operate at higher frequencies.

Ultraviolet recorders can achieve very fast recording speeds compared with other chart recorders, up to several metres per second. However, the light sensitive paper needs special storage and handling precautions to prevent it being damaged, and it is expensive.

Multichannel ultraviolet recorders are available, which simply use more moving coil mechanisms with the ultraviolet beam focused onto the same roll of chart paper.

Because of their sensitivity, high speed capability, and high running costs, ultraviolet recorders tend to be used more for detailed process analysis and research purposes rather than everyday quality monitoring.

Thermal array recorder

The thermal array recorder records data using semiconductor technology. Because of this, many models integrate signal conditioning, signal monitoring, and graphical output in one. Higher specification models can be considered as data acquisition systems, which are discussed in the next section.

Figure 8.24 shows a flow diagram of a basic thermal array recorder. It consists of an analogue to digital converter, which converts the analogue signal from the sensor into digital form. This digital signal is input to a print head controller which drives an array of several thousand thermal elements (the print head). The appropriate marker is triggered by a digital signal from the print head controller, and leaves a small mark on the thermally sensitive paper (as used in many fax machines). The paper spool is driven by a servo motor, and the chart speed and size of the axis are usually controlled manually

Thermal array recorders are accurate and versatile, and provide high quality images. Multiple channel devices are in common use, and some devices have memory allowing storage and retrieval of data. The paper is inexpensive and does not need the special storage requirements of ultraviolet sensitive paper.

Computers and data acquisition systems

Semiconductor technology has now developed to the point where recording and display of measured values and signals is usually achieved with the aid of a computer.

The term data acquisition refers to any process where information is converted into a form that can be handled by a computer. A computer-based data acquisition system is a system where a parameter, or parameters, are detected by a sensor, the output signal is then suitably conditioned, and this data is then stored or processed by a computer. Simple data acquisition systems may be referred to as data loggers.

Analogue measurement information is converted to digital data and fed into a computer for storage, real-time display, or recording for later analysis. All analogue and digital instruments can, if necessary, be simulated on a computer. When a computer simulates say, a voltmeter, it is known as a virtual voltmeter. The computer can simulate any type of instrument as long as the raw data is available at its inputs. This is convenient, and can provide a cost saving on the individual meters it replaces. This approach is useful, for example, in maintaining and controlling continuous industrial processes. Bar graphs, digital displays, and animated process simulations can show set values, process variables, error signals, value output and so on, to provide easily interpreted status information to the operator. Computers can quickly manipulate data, for example multiplying separate voltage and current signals together and displaying them in terms of power and energy. High quality hard-copies of graphs, charts, raw data, or other illustrative analysis can simply be printed out on a suitable peripheral printer or plotter.

> **Key fact**
>
> A multiplexer is a switch which routes information from several sources to one common destination.

Figure 8.25 is a flow diagram of a data acquisition system. Data can be obtained from a range of sensors by the one acquisition system. To achieve this a multiplexer is used. A multiplexer looks at (samples) each of the measurements (channels) in turn and feeds the sampled data to an analogue to digital (A-D) converter.

This converts the data into a form in which the computer can understand. Once the computer has received the data there are several options available. For example, it may manipulate the data appropriately and display it to the operator, or save it in its memory awaiting further data or for later use, it may be saved on disk, or fed back as part of a closed loop system.

The sample-and-hold device is necessary to ensure the sampled value is held constant during the time needed for conversion by the A-D converter. Multiplexers with the capacity to handle 64 channels are commonly available. Some A-D converters can readily convert in microseconds, allowing signals of up to hundreds of kilohertz to be sampled and recorded without difficulty.

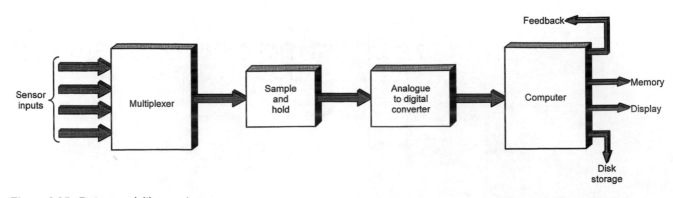

Figure 8.25 Data acquisition system

The computer-based data acquisition and analysis system has considerable flexibility. The range and nature of analysis and display of measured data is largely related to the features of the software application packages used. Huge amounts of data may be stored, allowing analysis of the behaviour of a parameter over time. Stored data can provide a trend display of a particular process variable, over, say, the previous hour. This facility is useful for quality control and system maintenance. Much of the human work involved is reduced, thus lessening the chances of human error. Computers can quickly and accurately manipulate data thus removing the need for long, repetitive or complex, time consuming calculations to be performed manually.

Although computers can do almost anything that other display and recording techniques are able to, they can be significantly more expensive. They cannot always be made in the small size of some other devices, and are not particularly robust. They need protection from dirty or dusty environments, and irreversible damage may be incurred by, for example, exposure to water or oil, or strong magnetic fields.

Summary

This chapter has introduced you to a selection of devices used to display and record values which sensors and transducers produce. There are many ways of achieving a visual representation of measurement and of producing a permanent record. The choice of techniques and equipment depends very much on the rate of signal variation with time and the detail and accuracy required.

Because of the development in computers and information technology, recording, analysis and display now nearly always uses digital techniques for industrial instrumentation and control applications.

Questions for further discussion

1. Some types of voltmeter and ammeter have both analogue and digital displays. What advantages does a meter with both types of display have?
2. A common cause of error in analogue devices which indicate values by means of a pointer over a scale (as in Figure 8.3) is known as parallax error. Explain what you think parallax error is, and ways in which it may be overcome.
3. A bar tube display (not discussed previously) basically consists of a group of seven fluorescent tubes which are selected to form characters, similar to the seven-segment display (if all are lit the display will show an 8). Give reasons why you think this type of display is now rarely used.
4. An ultrasonic flow meter is to be used to measure flow along an above ground water pipe. It needs to be used under varying conditions of weather and light, needs its own power supply, and has to be readily portable. What type of display or recording device would be most suited for use with this instrument? Give reasons for your answer and state any special precautions you would take.
5. Computer based oscilloscopes are available which have all the features of a standard cathode ray oscilloscope. Information is displayed on a personal computer monitor rather than viewed on an oscilloscope cathode ray tube. What advantages do you think a computer based oscilloscope has over the standard type? Why, in some situations, may you prefer to use a standard cathode ray oscilloscope?

9 Signal Conditioning and Interfacing: Passive Circuit Techniques

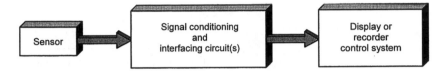

Figure 9.1 Matching sensor to load

Most sensors intended for remote reading produce an electrical output signal. For these signals to be of use, they have to be compatible with the inputs to the appropriate display, recorder or control system. Direct compatibility is rarely found and appropriate methods are needed to modify signals and electrically match system elements to each other. Modifying signals is known as signal conditioning and matching system devices electrically is known as interfacing. Figure 9.1 shows this in flow diagram form.

The range of sensors, displays and recorders considered in Chapters 3 to Chapter 8 give rise to a wide variety of signal conditioning and interfacing requirements. For example, the output signal from a sensor may need scaling down, or a change in resistance may need calibrating in terms of a voltage. To obtain the best performance from a display or recording device, it is often necessary to transfer the maximum power or signal voltage from the sensor. In many cases purely passive techniques are adequate. Passive circuits obtain their power from the system they are connected to.

Key fact

A passive circuit is an electrical circuit which employs only passive components, that is, resistors, capacitors, or inductors. Passive components do not have their own source of power.

In this chapter we shall concentrate on commonly used passive circuit techniques. An understanding of electrical theory is required, so if you are in any doubt about your knowledge of basic d.c. and a.c. theory you should revise these subjects before proceeding with this chapter, or Chapter 10. By its nature, many of the proofs are mathematical.

Signal conditioning

In this section we will look at how the potentiometer and Wheatstone bridge condition signals. We have previously met both of these devices. However, in the case of the potentiometer we have not yet looked at its use in signal conditioning, and we have only briefly looked at the function of the Wheatstone bridge.

The basic principles of both devices are explained by studying the circuits used and looking at the associated mathematics.

Passive voltage division: the potentiometer

We met two types of potentiometer, the linear potentiometer and the angular potentiometer, in Chapter 3. However, potentiometers have uses other than measuring displacement. They can also be used to condition output signals from other sensors.

A potentiometer is a device that allows a lower voltage to be derived from a higher voltage. The output is a scaled-down, or divided, version of the input. If the output from a sensor is too high for the specification of the display or recording device, we can use a potentiometer to scale down (reduce) the voltage accordingly. If the amount by which the signal has been scaled down is known, we can calibrate the displayed or recorded values by this factor.

For this we need a method of calculating the output voltage V_o.

Key fact

A potential (or voltage) divider is a chain of resistances in series, such that the voltage across one or more is an accurately known fraction of the voltage applied to all the others.

Figure 9.2 shows the basic potentiometer circuit, where V_i is the input voltage, V_o is the output voltage, and R is the resistance.

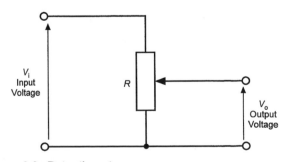

Figure 9.2 Potentiometer

To calculate the scaling factor we divide the output voltage by the input voltage, which is

$$\frac{V_o}{V_i}$$

To allow us to represent the position of the sliding contact in more detail, we can redraw the potentiometer to the circuit shown in Figure 9.3. This divides the resistance R into the components R_o and R_1, which have corresponding voltages V_{Ro} and V_{R1}. R_o is the resistance of the potentiometer resistance element across the output R_1 is the remaining resistance of the potentiometer. Varying the position of the potentiometer sliding contact varies the values of R_o and R_1 and hence the output voltage V_o.

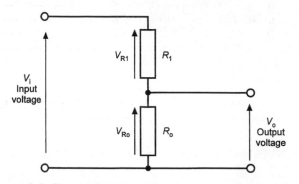

Figure 9.3 Potential divider

If no current flows to the output, then the current I supplied by the input voltage V_i will be

$$I = \frac{V_i}{R_1 + R_o}$$

The voltages across resistors R_o and R_1, will be $V_{R1} = IR_1$ and $V_{Ro} = IR_o$ respectively.

From Kirchhoff's Voltage Law we know that $V_i = V_{R1} + V_{Ro}$, and since $V_o = IR_o$, we have

$$\frac{V_o}{V_i} = \frac{IR_o}{IR_1 + IR_o}$$

Because the value of I is constant and is common to the denominator and numerator in the right-hand side of the equation, we can remove it. This gives,

$$\frac{V_o}{V_i} = \frac{R_o}{R_1 + R_o}$$

Rearranging this with respect to V_o gives

$$V_o = V_i\left(\frac{R_o}{R_1 + R_o}\right)$$

Thus by knowing the resistances of the potentiometer and the input voltage we can determine the output voltage. We can then determine the scaling factor. Figure 9.4 is a sample calculation determining the scaling factor of an unloaded potentiometer.

Key fact

Kirchhoff's Voltage Law states that the sum of the voltage drops around any closed circuit is zero.

It is important to note that this proof assumes that no current is taken by the circuit to which V_o is connected. If the potentiometer is allowed to act as a current source, that is, a resistance is connected across it, then the relationship

$$V_o = V_i\left(\frac{R_o}{R_1 + R_o}\right)$$

will no longer be true. Also, the measured output voltage V_o will be reduced. This reduction is directly related to the amount of current diverted away from R_1 in the change in circuit parameters. In practice, any display or recording device is likely to introduce a resistance across the potentiometer. It is necessary to reduce the effect of this resistance on the scaling factor of the potentiometer as much as possible.

Problem

Consider the potential divider circuit diagram shown below. The input voltage is 12 volts, and the sliding contact is set such that the resistance across the output voltage is 8 kΩ while the remainder of the resistance element gives a resistance of 2 kΩ. Determine the output voltage and hence the scaling factor of the potentiometer at this setting.

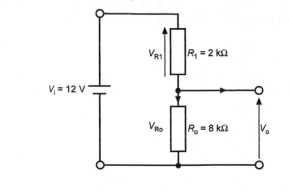

Solution

From the question, we know that:
$V_i = 12$ volts, $R_o = 8$ kΩ and $R_1 = 2$ kΩ

$$V_o = V_i\left(\frac{R_o}{R_1 + R_o}\right)$$

$$= 12\left(\frac{8}{2+8}\right) = 12 \times 0.8$$

$$\therefore V_o = \mathbf{9.6\ V}$$

$$\text{Scaling factor} = \frac{\text{output voltage}}{\text{input voltage}} = \frac{V_o}{V_i}$$

$$= \frac{9.6}{12}$$

Scaling factor = 0.8 or 80%

Figure 9.4 Sample calculation determining the scaling factor of an unloaded potentiometer

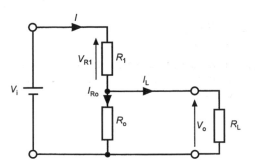

Figure 9.5 Potentiometer circuit with a load resistance

Figure 9.5 shows a potentiometer circuit with a load resistance across it. Some current in the circuit will be diverted and flow through this resistance. If the amount of current which flows through the load resistance is reduced, we can reduce its effect on the scaling factor.

To calculate the effect of the load resistance, we first need to determine the equivalent resistance, R_E, of the parallel combination of the load resistance R_L, and R_o which effects V_o. This is given by the expression,

$$R_E = \frac{R_o \times R_L}{R_o + R_L}$$

which is the standard expression for two resistors in parallel.

To find the output voltage, we replace R_1 with R_E, giving the equation to find the output voltage of a potentiometer circuit with a load resistance as

$$V_o = V_i\left(\frac{R_o}{R_E + R_o}\right)$$

The larger the load resistance R_L, the less current will flow through it and hence the less the effect on output voltage, or the scaling factor. The example calculation given in Figure 9.6 illustrates this.

Figure 9.6 shows that when the load increases, the loading error reduces. When the smaller 3 kΩ load resistance is connected across the potentiometer the output voltage is 1.3 volts, which is a decrease of 0.7 volts from the unloaded circuit, or a difference of 35%. However, when the larger 200 kΩ resistance is connected, the decrease in the output voltage from the unloaded circuit is only 0.02 volts, a difference of only 1%.

In practice, the load resistance should be at least 100 times that of R_o to ensure there is no more than 1% error.

Problem

Consider the two circuit diagrams shown below. The only difference between the two circuits is that the load resistance of circuit (a) is 3 kΩ, while the load resistance of circuit (b) is 200 kΩ. For both circuits, the input voltage is 10 volts, and the sliding contact of the potentiometer is set such that the resistance across the output voltage is 2 kΩ while the remainder of the resistance element gives a resistance of 8 kΩ.

Determine the output voltage when there is no load on the circuit, the output voltage when the load resistance is 3 kΩ, and the output voltage when the load resistance is 200 kΩ.

Solution

From the question, we know that:
V_i =10 volts, R_o = 2 kΩ and R_1 = 8 kΩ

At no load,

$$V_o = V_i\left(\frac{R_o}{R_1+R_o}\right) = 10\left(\frac{2}{8+2}\right) = 10 \times 0.2$$

$$V_o = \mathbf{2\,V}$$

When the load resistance is 3 kΩ,

$$R_E = \frac{R_o \times R_L}{R_o + R_L} = \frac{2 \times 3}{2+3} = \frac{6}{5}$$

$$R_E = 1.2\,\Omega$$

$$V_o = V_i\left(\frac{R_E}{R_1+R_E}\right) = 10\left(\frac{1.2}{8+1.2}\right) = 10 \times 0.130$$

$$V_o = \mathbf{1.3\,V}$$

When the load resistance is 200 kΩ,

$$R_E = \frac{R_o \times R_L}{R_o + R_L} = \frac{2 \times 200}{2+200} = \frac{400}{202}$$

$$R_E = 1.98$$

$$V_o = V_i\left(\frac{R_E}{R_1+R_E}\right) = 10\left(\frac{1.98}{8+1.98}\right) = 10 \times 0.198$$

$$V_o = \mathbf{1.98\,V}$$

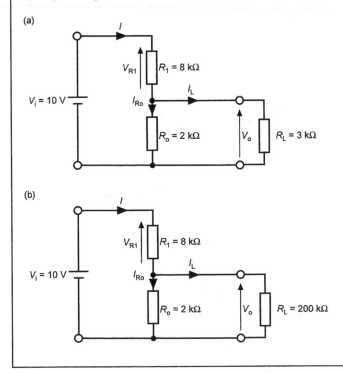

Figure 9.6 Effect of load resistances on the output voltage of a potentiometer circuit

We shall see in Chapter 10 that techniques are available to present an almost infinite load resistance to the circuit feeding it. It is not unusual to have voltmeters with input impedances greater than 1 MΩ. As a result, a high level of accuracy can be achieved since loading effects can be negligible.

Wheatstone bridge

Previously we have seen that several sensors measure parameters in terms of a change in resistance. We met the Wheatstone bridge circuit in previous chapters, used with bonded resistance strain gauges and strain gauge load cells. It is also used with devices such as metal resistance thermometers and thermistors. All these devices need to have their change in resistance calibrated with respect to the parameter they are measuring. When the resistance of a circuit changes, the current and voltage also change proportionally. Hence the Wheatstone bridge is a device commonly used to condition signals with respect to voltage or current as well as resistance.

Key fact

The Wheatstone bridge, named after Sir Charles Wheatstone (1802–1875), is a device used to determine the value of an unknown resistance by comparing it with a known resistance

---

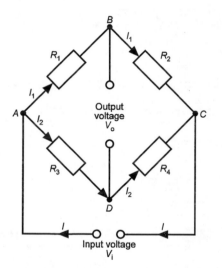

Figure 9.7 Wheatstone bridge

Figure 9.7 shows a Wheatstone bridge circuit. Usually one of the resistors is a sensor. For example, R_1 may be a metal resistance thermometer. Because its resistance changes with temperature, its resistance is unknown. Similarly it may be a bonded resistance strain gauge, whose resistance changes when it is under a strain.

The Wheatstone bridge circuit shown in Figure 9.7 can be identified as comprising of two potential dividers. One is formed by resistors R_1 and R_2, and the other by R_3 and R_4.

The Wheatstone bridge is of particular use as a signal conditioning system because we can establish a linear relationship between the output voltage V_o and a change in resistance of one of the resistors. To establish this relationship, we can consider the Wheatstone bridge as two potentiometers in parallel. When the output voltage V_o is zero then the potential at B is equal to the potential at D. Thus,

$$V_{R1} = V_{R3}$$

giving,

$$I_1 R_1 = I_2 R_3 \qquad [1]$$

Similarly,

$$V_{R2} = V_{R4}$$

so

$$I_1 R_2 = I_2 R_4 \qquad [2]$$

Dividing [2] into [1],

$$\frac{I_1 R_1}{I_1 R_2} = \frac{I_2 R_3}{I_2 R_4}$$

$$\frac{R_1}{R_2} = \frac{R_3}{R_4}$$

The input voltage V_i is connected between points A and C. To relate this to the output voltage V_o, we can see that

$$V_o = V_{R1} - V_{R3}$$

where:

$$V_{R1} = I_1 R_1, \text{ and } V_{R3} = I_2 R_3$$

If $I_1 = \dfrac{V_i}{R_1 + R_2}$ and $I_2 = \dfrac{V_i}{R_3 + R_4}$

then

$$V_{R1} = \left(\frac{V_i}{R_1 + R_2}\right) R_1$$

$$\therefore V_{R1} = \frac{V_i R_1}{R_1 + R_2}$$

and

$$V_{R3} = \left(\frac{V_i}{R_3 + R_4}\right) R_3$$

$$\therefore V_{R3} = \frac{V_i R_3}{R_3 + R_4}$$

$$V_o = V_{R1} - V_{R3}$$

$$\therefore V_o = V_i\left(\frac{R_1}{R_1 + R_2} - \frac{R_3}{R_3 + R_4}\right) \qquad [3]$$

Equation [3] is an equation to determine V_o. However, if the unknown resistance is R_1 then the relationship between R_1 and output voltage V_o in equation [3] is not a linear relationship. For the purposes of signal conditioning it would be far more useful to have an expression for V_o which gives a linear relationship with R_1, the unknown resistance.

To determine a linear relationship, suppose resistance R_1 changes by ΔR_1, to $R_1 + \Delta R_1$, giving an output voltage change by ΔV_o. The total output voltage is then $V_o + \Delta V_o$. The equation will then become,

$$\therefore V_o + \Delta V_o = V_i\left(\frac{R_1 + \Delta R_1}{R_1 + \Delta R_1 + R_2} - \frac{R_3}{R_3 + R_4}\right) \qquad [4]$$

To determine ΔV_o, subtract equation [3] from equation [4], to give

$$\Delta V_o = V_i \left(\frac{R_1 + \Delta R_1}{R_1 + \Delta R_1 + R_2} - \frac{R_1}{R_1 + R_2} \right)$$

Now, if ΔR_1 is much less than R_1, then we can approximate the value of the change in output voltage ΔV_o to

$$\Delta V_o \approx V_i \left(\frac{R_1 + \Delta R_1}{R_1 + R_2} - \frac{R_1}{R_1 + R_2} \right)$$

$$\therefore \Delta V_o \approx V_i \left(\frac{\Delta R_1}{R_1 + R_2} \right)$$

This is a linear relationship between ΔV_o and ΔR_1. Therefore, as long as ΔR_1 is much less than R_1, and the display or recorder circuit supplied by V_o takes very little current compared to that supplied to the Wheatstone bridge, we can assume that ΔV_o is directly proportional to ΔR_1. This makes the Wheatstone bridge a very useful device for providing a conditioned output signal to a display or recording device for sensors based on a change in resistance.

Figure 9.8 is an example calculation of conditioning a signal from a platinum resistance thermometer (see Chapter 6) using the Wheatstone bridge. Similar calculations can be made for any application based on a change in resistance.

Passive interfacing

In previous chapters we discussed the properties of resistance and capacitance. In this section we come across the property of electrical inductance. A device which has inductance is known as an inductor. An inductor is a device in an electrical circuit which, when carrying current, forms a magnetic field and stores magnetic energy.

Various techniques are used to ensure that the maximum signal or the maximum power is transferred from a sensor to the display or recording device. The term matching is used, which means that the optimum ratio of parameters (such as resistance) exists between one side of the circuit to the other. As we shall see, this ratio is not necessarily equal, or 1:1, as implied by the term matching.

The basic principles of signal matching for maximum signal voltage transfer, power matching for maximum power transfer, and impedance matching for maximum power transfer using transformers are discussed in this section. As in the previous section discussing signal conditioning, these techniques are explained by studying the circuits used and looking at the associated mathematics.

Problem

Consider a platinum resistance thermometer connected as one arm of a Wheatstone bridge, as shown in the diagram below. The fixed resistances in the circuit are 200 Ω each, and the input voltage is 10 V. At 0 °C the circuit is balanced and at this temperature the thermometer has a resistance of 200 Ω.

If the temperature coefficient of resistance of platinum is 0.004 °C^{-1}, determine the change in output voltage for a 1 °C change in temperature.

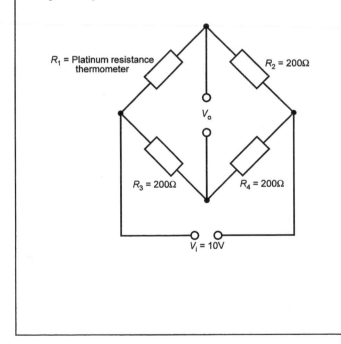

Solution

From Chapter 6, we know that

$$R_t = R_0 (1 + \alpha t)$$

where:

- R_0 is the resistance in ohms of the conductor at a temperature of 0 °C.
- R_t is the resistance in ohms of the conductor at t °C.
- α is the temperature coefficient of resistance of the material.

From the question, we know that R_0 is 200 Ω, α is 0.004 °C^{-1}, and t is 1 °C. Therefore,

$$R_t = 200 \times \left[1 + \left(0.004 \times 1 \right) \right] = 200 \times 1.004$$

$$R_t = 200.8 \ \Omega$$

$$\Delta R = R_t - R_0$$

$$\Delta R = 0.8 \ \Omega$$

Because ΔR is much less than R_0, we can use the expression

$$\Delta V_o \approx V_i \left(\frac{\Delta R_1}{R_1 + R_2} \right)$$

$$\Delta V_o \approx 10 \left(\frac{0.8}{200 + 200} \right) = 10 \times 0.002$$

$$\Delta V_o \approx \textbf{0.02 volts}$$

Thus, for each degree change in temperature, there is a 20 mV change in the output voltage.

Figure 9.8   Sample calculation of a Wheatstone bridge circuit based on a platinum resistance thermometer

Power matching: maximum power transfer

It is often desirable, if not essential, that the maximum amount of energy is delivered from a source to a load. This will be the case, for example, in applications in which the current supplied is required to produce a magnetic effect, such as in moving coil meters, or when the output of a control system is fed to say, an electrically operated valve controlling fluid delivery in a chemical plant. In cases where the energy delivered from a source to a load must be maximised, power matching is necessary. Power matching is achieved when the source resistance is the same as the load resistance. If the source has an e.m.f. E_s and internal resistance R_s the circuit is as shown in Figure 9.9.

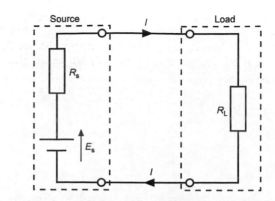

Figure 9.9 Load supplied from a power source

Problem

In the circuit below, a variable load resistor is connected in series with a power source. The power source has an e.m.f. of 50 V, and an internal resistance of 25 Ω.

Calculate the power dissipated in the load resistor when its value is varied in increments of 5 Ω from 0 Ω to 50 Ω. Draw a graph of your results and determine the maximum power dissipated and the resistance of the load resistor at this value.

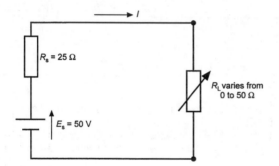

Solution

From the question, we know that E_s is 50 V, and R_s is 25 Ω. R_L increases in steps of 5 Ω from 0 Ω to 50 Ω. The power P_L dissipated in the load resistor will be,

$$\text{Power } P_L = I^2 \times R_L$$

where: $I = \dfrac{E_s}{R_T}$ and $R_T = R_L + R_s$

Using these equations, the results can be tabulated, and from these the graph of power dissipated against load resistance plotted, as shown below:

R_L (Ω)	$R_T = (R_L + R_s)$ (Ω)	$I = E_s/R_T$ (A)	$P_L = I^2 . R_L$ (W)
0	25	2.00	0
5	30	1.67	13.89
10	35	1.43	20.49
15	40	1.25	23.43
20	45	1.11	24.69
25	50	1.00	25.00
30	55	0.91	24.79
35	60	0.83	24.31
40	65	0.77	23.67
45	70	0.71	22.96
50	75	0.67	22.22

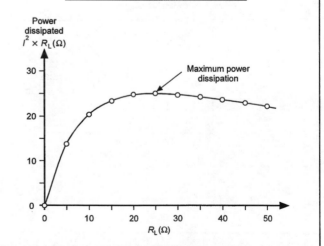

From both the table and the graph, the maximum power dissipated is **25 watts** at which **the load resistance is equal to the source resistance**, at **25 ohms**.

Figure 9.10 Sample calculation showing maximum power transfer

The power transferred from the source to the load is a maximum, and the load is said to be matched to the source, when the resistance of the load R_L is equal to the internal resistance R_s of the source. This is illustrated by the example calculation given in Figure 9.10.

The calculations show that a maximum power dissipation of 25 watts occurs when $R_L = R_s$ which in this case is 25 Ω. At all other values of R_L the power dissipated by R_L is less than 25 watts.

Figure 9.10 considered a purely resistive power source and a purely resistive load. In practice sources and loads are unlikely to be purely resistive, there will also be capacitive or inductive opposition to current flow (reactance). In these cases, source resistance must be equal to the load resistance and also the source reactance must be balanced by a reactance of the same magnitude but opposite type. That is, in order to achieve maximum power transfer, if the load is inductive, the source should be capacitive.

Signal matching: maximum voltage transfer

Output signals produced by sensors are usually small. Because of this it is very important that as much as possible of the signal voltage is transferred to the display, recorder or control system. To achieve this, as we saw with the potentiometer and the Wheatstone bridge circuit, the resistance of the load should be much higher than the resistance of the signal source.

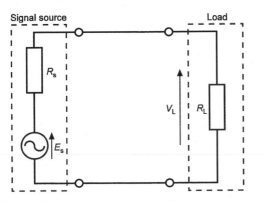

Figure 9.11 Signal source connected to a load

Figure 9.11 shows a signal source from a sensor, connected to a load, such as a display. The signal source has internal e.m.f. E_s and resistance R_s. If current is supplied, the voltage of the load R_L is less than E_s.

The resistance of the source and load act as a potential divider. The effective voltage across the load, V_L, depends upon the ratio of R_s and R_L. So, from Kirchhoff's Law:

$$E_s - IR_s - IR_L = 0$$

$$\therefore E_s = IR_L - IR_s$$

$$E_s = I(R_L - R_s)$$

If $R_L \gg R_s$, then

$$E_s = IR_L = V_L$$

That is, there is no loss.

If R_L is very much greater than R_s then the difference between the signal voltage and the voltage across the load is relatively small. In this case signal transfer has been maximised.

Impedance matching using transformers

The purpose of impedance matching using transformers is to ensure that the maximum power is transferred from the sensor to the display or recording device. Because of the nature of transformers this only works for a.c.

Previously in this book we have seen that transformers are devices used to change voltages in a circuit. Similarly transformers can be used to change impedances within a circuit.

> **Key fact**
>
> Impedance is a measure (in ohms) of opposition to flow of a.c. current by resistance and reactance

An advantage of the transformer is that it acts as a buffer between the sensor and display or recording device. The voltage across the secondary winding can be considerably different to that of the source. Power is transferred via the magnetic field linking the transformer windings and, since there is no electrical connection between the primary and secondary windings, the device is inherently safe.

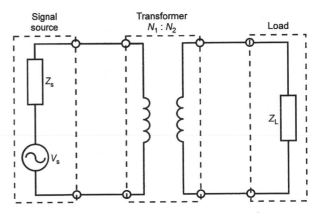

Figure 9.12 Impedance matching using a transformer

Figure 9.12 shows a circuit using a transformer to match the impedance of a sensor (the source), Z_s, to the impedance of a display or recording device (the load) Z_L, and hence achieve maximum power transfer. The number of windings on the primary coil of the transformer is N_1, and the number of windings on the secondary coil is N_2. To match the impedances, an appropriate ratio of transformer primary windings to secondary windings must be used.

It can be shown that,

$$\frac{N_1}{N_2} = \sqrt{\frac{Z_s}{Z_L}}$$

That is, to achieve maximum power transfer from the source to the load, the ratio of the transformer primary to secondary windings must be equal to the square root of the source impedance divided by the load impedance.

An example calculation is given in Figure 9.13.

Problem

In the circuit below, a signal from a sensor is being transferred to a display. The impedance of the sensor is 200 kΩ, while the impedance of the display is much less, at 2 kΩ. Assuming transformer losses are negligible, determine the turns ratio of the transformer so the impedance of the sensor is matched to the display, thus ensuring maximum power transfer.

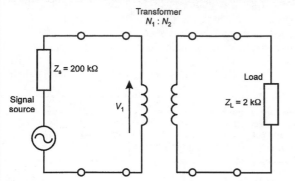

If the voltage across the primary winding is 50 V when maximum power transfer occurs, determine the power delivered to the transformer.

Solution

From the question, we know that Z_s = 200 kΩ, Z_L = 2 kΩ, and at maximum power transfer V_1 = 50 V. To determine the turns ratio of the transformer,

$$\frac{N_1}{N_2} = \sqrt{\frac{Z_s}{Z_L}} = \sqrt{\frac{200}{2}} = \sqrt{100}$$

$$\frac{N_1}{N_2} = 10$$

Therefore the turns ratio of the transformer for maximum power transfer is 10:1. At maximum power transfer, $Z_s = Z_1$ = 200 kΩ. To determine P_1, the power delivered to the transformer,

$$P_1 = \frac{V_1^2}{Z_1} = \frac{50^2}{200} = \frac{2500}{200}$$

$$P_1 = 12.5 \text{ Watts}$$

Figure 9.13 Impedance matching using transformers, example calculation

Summary

This chapter has considered a range of passive techniques commonly used to modify or condition sensor outputs, to make them compatible with displays, recorders or control elements. You will have noted that practical performance is usually somewhat different to the theoretical and care must be taken with loading effects.

Most applications you will meet will use a combination of the passive techniques you have studied in this chapter and the active methods that you will meet next in Chapter 10.

Questions for further discussion

1. The sliding contact of the potentiometer circuit in Figure 9.4 is adjusted so that the resistance across the output voltage is 4 kΩ while the remainder of the resistance element gives a resistance of 6 kΩ. Determine the scaling factor of the potentiometer at this new setting (the input voltage remains at 12 V).

2. Consider the Wheatstone bridge circuit with a platinum resistance thermometer shown in Figure 9.8. If the input voltage to this circuit was doubled to 20 volts, what effect would this have on the output voltage for each degree change in temperature? What advantages may there be in using a larger input voltage? Are there any disadvantages?

3. Consider the circuit in Figure 9.9, showing a signal source connected to a load. If V_s is 10 V, R_L is 200 kΩ, and R_s is 2 kΩ, calculate (to two decimal places) the difference between the values of voltage across the source V_s and voltage across the load V_L. Repeat the calculation assuming R_L has reduced to 20 kΩ. What conclusions do you draw from these answers?.

4. Consider the circuit shown in Figure 9.10. If the source resistance R_s is changed to 30 kΩ, determine the power dissipated in the load resister R_L when maximum power is being transferred.

5. Calculate the turns ratio required by the transformer to match the impedances in the circuit shown in Figure 9.14, if the impedance of the source Z_s is changed to 180 kΩ and the impedance of the load is Z_L changed to 20 kΩ. At maximum power transfer, determine the power delivered to the transformer.

6. In practice, losses occur within transformers. What do you think the nature of these losses are, and what effect would they have on the equations we used for impedance matching using transformers?

10 Signal Conditioning and Interfacing: Active Circuit Applications

Passive techniques alone cannot satisfy all sensor signal conditioning and interfacing requirements. In practice, the passive techniques we looked at in Chapter 9 are usually used with active devices.

> **Key fact**
>
> Active devices are electrical components such as diodes, transistors, and integrated circuits which can control voltages and currents. Therefore they can change the magnitude of the voltage or current, or produce switching action in a circuit.

An amplifier is an electronic device, or group of devices, which increase the size of (amplifies) a voltage or current signal, without altering the signal's basic characteristics. It is made up of active and passive components, and has a power supply separate from the signal it is acting on.

Operational amplifiers are a special type of amplifier. They are termed 'operational' because they were originally developed for early computers to perform basic mathematical operations such as adding and subtracting. They are the basic building blocks of most active electronic signal conditioning circuits. In the form of integrated circuits, they are relatively inexpensive, precise, and reliable.

Operational amplifier techniques are probably the most commonly used active method of signal conditioning. In this chapter we will look at how they are included in electronic circuits to perform various interfacing and signal conditioning roles.

Operational amplifier circuits

Figure 10.1 shows the schematic diagram of an operational amplifier. The internal detail of the integrated circuits which make up operational amplifiers consists of a complex arrangement of transistors and resistors. However, it is the function of the integrated circuit as a whole which the user needs to know, and this is how they are represented here.

Operational amplifiers are often referred to in their abbreviated form as 'op amps'.

Operational amplifiers have two inputs and one output. The two inputs in Figure 10.1 are marked V_1 and V_2. The input V_1 is indicated by a '−' sign. The minus sign shows that the value of this input is inverted. The input V_2 is indicated by a '+' sign, and is not inverted. Thus the amplifier subtracts the value of V_1 from V_2, and the output signal is an amplified version of the difference between the two input signals.

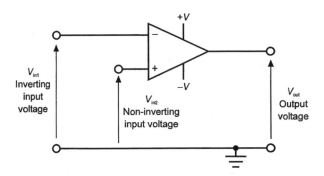

Figure 10.1 Operational amplifier

The supply to the operational amplifier is shown in Figure 10.1 as $+V$ and $-V$. Diagrams normally assume the supply to be present and do not show it.

> **Key fact**
>
> The gain of an amplifier is the amount by which it increases the size of a signal, that is, the ratio of the output signal to the input signal. If the gain is greater than 1 the output will be bigger than the input.

One of the important properties of operational amplifiers is that their gain is very high. A typical gain value is 50000, which is enough for most practical purposes.

The input impedance of an ideal or perfect operational amplifier would be infinite. This is not possible in practice, but the input impedance is usually very high, for example 1 MΩ. If the input resistance of the external circuits is made small relative to the input impedance of the amplifier, the amplifier will not draw significant current through the external resistors.

Similarly, the output impedance of an ideal operational amplifier would be zero. Again this is not possible in practice but the output impedance of an operational amplifier can be as low as 100 Ω. If the resistance of the load is made high relative to the output impedance of the amplifier, the amplifier will again approximate to the ideal.

The general equation for the output voltage V_{out} of an operational amplifier is:

$$V_{out} = A_o(V_{in2} - V_{in1}) \text{ volts}$$

where:

- V_{out} is the output voltage
- A_o is the open loop gain of the amplifier
- V_{in1} is the inverting input voltage
- V_{in2} is the non-inverting input voltage

Sensors and measurement and control circuits use operational amplifiers in several different arrangements. Operational amplifiers, along with other devices, both passive and active, are used to build signal conditioning circuits with varying functions. In the following sections we shall discuss the configurations and operation of several of these circuits.

Voltage comparator

The basic operational amplifier can act as a voltage comparator. Figure 10.2 shows an operational amplifier voltage comparator circuit.

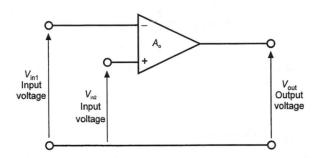

Figure 10.2 Voltage comparator

The circuit acts as a differential amplifier which compares the two input voltages V_{in1} and V_{in2}. It produces an output with a polarity which indicates which of the inputs is the greater.

$$V_{out} = A_o(V_{in2} - V_{in1})$$

- If V_{in1} is greater than V_{in2}, then V_{out} is negative.
- If V_{in1} is less than V_{in2}, then V_{out} is positive.
- If V_{in1} is the same as V_{in2}, then V_{out} is zero.

With a gain of 50000 and supply of ±5 V, a difference of 0.1 mV is sufficient to bring the output to saturation point (it cannot go any higher or lower). In practice this means that if the input passes through zero in either direction, the output reaches its maximum at the supply value. This is shown in Figure 10.3.

If a sine wave is applied to V_{in2} and V_{in1} is connected to 0 V, V_{out} will change state each time the sine wave passes through zero. The output waveform will be a square wave. Figure 10.3 shows this.

This type of comparator is often referred to as a zero-crossing detector.

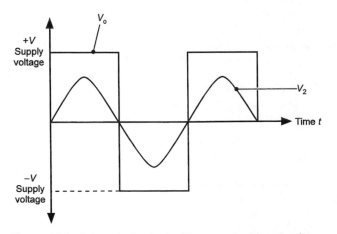

Figure 10.3 Integrated output with a constant input voltage

Inverting amplifier

Figure 10.4 shows an inverting operational amplifier. The input voltage is connected to the inverting terminal of the amplifier via resistor R_i. With this arrangement, the output voltage V_{out} is an inverted form of the input voltage V_{in}. The amplitude of V_{out} and V_{in} will be identical if the feedback resistor R_f has the same value as R_i.

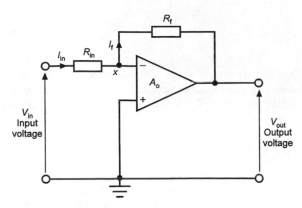

Figure 10.4 Inverting amplifier

Point x is a virtual earth. This means that it has the same potential difference as the Earth, which is usually considered to be 0 V. This means that:

$$I_{in} = \frac{V_{in}}{R_{in}}, \text{ and } I_f = \frac{V_{out}}{R_f}$$

so,

$$\frac{V_{in}}{R_{in}} = -\frac{V_{out}}{R_f}$$

Hence

$$\frac{V_{out}}{V_{in}} = -\frac{R_f}{R_{in}}$$

Because the gain of the circuit A is

$$A = \frac{V_{out}}{V_{in}}$$

The voltage gain A for an inverting operational amplifier is

$$A = -\frac{R_f}{R_{in}}$$

where the negative sign indicates the inversion of the input voltage. Figure 10.5 is an example calculation to determine the output voltage of an inverter amplifier.

Figure 10.6 shows a non-inverting operational amplifier. With this arrangement, the output voltage V_{out} is identical in polarity (in-phase) with the input voltage V_{in}. The value of V_{out} and two times V_{in} if the feedback resistor R_f has the same value as the resistance on the input R_{in}.

Problem

Consider the inverting operational amplifier shown opposite. The resistance on the input is 5 kΩ and the feedback resistance is 100 kΩ. If an input voltage of 50 mV is applied, determine the gain of the amplifier and hence determine the output voltage.

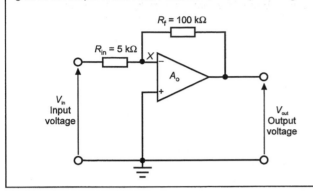

Solution

From the question we know that $R_i = 5$ kΩ, $R_f = 100$ kΩ, and $V_{in} = 50 \times 10^{-3}$ V

$$A = -\frac{R_f}{R_i} = \frac{-100 \times 10^3}{5 \times 10^3}$$

$$A = -20$$

If the gain of the amplifier is –20, then the output voltage is

$$V_{out} = 50 \times 10^{-3} \times -20 = -1$$

Therefore the output voltage is –1 V

Figure 10.5 Calculation determining the output voltage of an inverter amplifier

Non-inverting amplifier

Figure 10.6 shows a non-inverting amplifier circuit.

For the non-inverting operational amplifier circuit, $I_{in} = I_f$ and

$$I_{in} = \frac{V_{in}}{R_{in}}, \text{ and } I_f = \frac{V_{out}}{R_f + R_{in}}$$

so,

$$\frac{V_{in}}{R_{in}} = \frac{V_{out}}{R_f + R_{in}}$$

Rearranging

$$\frac{V_{out}}{V_{in}} = \frac{R_f + R_{in}}{R_{in}} = \frac{R_f}{R_{in}} + 1$$

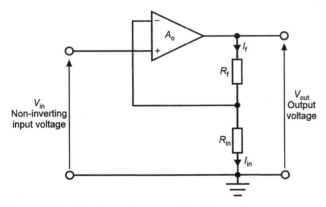

Figure 10.6 Non-inverting operational amplifier

Therefore, the voltage gain A of this non-inverting operational amplifier circuit is

$$A = 1 + \frac{R_f}{R_{in}}$$

The input voltage can be a.c. or d.c. The output voltage will lie between the positive and negative voltage levels of the amplifier supply.

Summing amplifier

The summing operational amplifier is an inverting amplifier with multiple inputs. It adds the values of two input voltages together, and inverts the value of this sum.

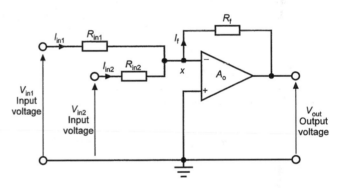

Figure 10.7 Summing amplifier

Figure 10.7 shows a summing operational amplifier. At point x,

$$I_{in1} + I_{in2} = I_f$$

$$\frac{V_{in1}}{R_{in1}} + \frac{V_{in2}}{R_{in2}} = \frac{-V_{out}}{R_f}$$

$$V_{out} = -R_f\left(\frac{V_{in1}}{R_{in1}} + \frac{V_{in2}}{R_{in2}}\right)$$

If $R_f = R_{in1} = R_{in2}$, then

$$V_{out} = -(V_{in1} + V_{in2})$$

Figure 10.8 shows an example of the use of a summing amplifier. The result in Figure 10.8 is an output voltage whose amplitude is the difference between the two signals, since they are opposite phase. The output voltage will be phase inverted with respect to the input signal with the larger amplitude.

Problem

Consider the summing operational amplifier shown below. The resistances on the inputs are both 10 kΩ, and the feedback resistance is also 10 kΩ. If the input signal voltages are 40 mV and −55 mV, determine the output voltage.

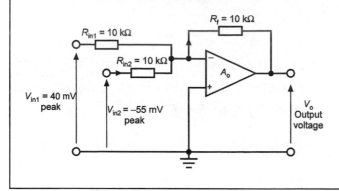

Solution

From the question we know that R_{in1} = 10 kΩ, R_{in2} = 10 kΩ, and R_f = 10 kΩ. V_{in1} = 40 × 10^{-3} V and V_{in2} = −55 × 10^{-3} V

$$V_{out} = -(V_{in1} + V_{in2})$$

$$= -(40 \times 10^{-3} - 55 \times 10^{-3}) = -(-15 \times 10^{-3})$$

$$V_{out} = 15 \times 10^{-3} \text{ V}$$

$$V_{out} = 15 \text{ mV}$$

Figure 10.8 Calculation determining the output voltage of a summing operational amplifier

Voltage follower

The voltage follower is used where it is important that sensor circuits are not placed under load. The operational amplifier is suited to this application because of its high input impedance and low output impedance.

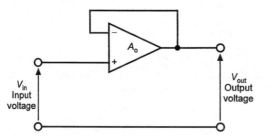

Figure 10.9 Voltage follower with unity gain

Figure 10.9 shows an operational amplifier as a non-inverting voltage follower. The gain of this amplifier is unity, which means that V_{out} is the same as V_{in}.

Differential amplifier

The differential amplifier is used to amplify the difference between two voltages, as produced by a Wheatstone bridge for example. Figure 10.10 shows the differential operational amplifier arrangement.

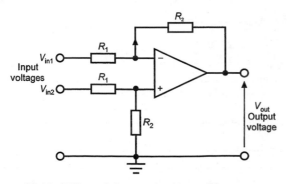

Figure 10.10 Differential operational amplifier

For a differential amplifier the output voltage is,

$$V_{out} = \left(\frac{R_2}{R_1}\right) \times (V_2 - V_1)$$

However, because the voltage dividers R_2 and R_1 are connected to the input, the effective input impedance is reduced. To compensate for this, the instrumentation amplifier was developed. Figure 10.11 shows one of the most popular configurations of instrumentation amplifier.

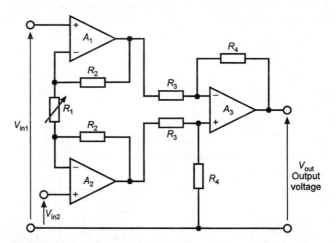

Figure 10.11 Instrumentation amplifier

In the instrumentation amplifier, the operational amplifiers A_1 and A_2 are voltage followers which provide a very high impedance input stage for the differential amplifier A_3. Variable resistance R_1 provides fine adjustment of the balance between A_1 and A_2. This rejects input signals of the same polarity, for example the d.c. levels at the two ends of the Wheatstone bridge output.

Integrating amplifier

The integrating amplifier is similar to the inverting amplifier except that the feedback circuit uses a capacitor (C_f) instead of a resistor. Figure 10.12 shows an integrating circuit.

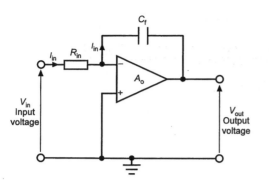

Figure 10.12 Integrating circuit

The term 'integrating' refers to the fact that the circuit can be used to add up all values of V_{in} over a period of time, t. In this case:

$$V_{out} = -\frac{1}{R_{in}C_f}\int V_{in}\ dt$$

For this circuit, the output voltage V_{out} is:

$$V_{out} = -\frac{I_{in}\times t}{C_f}$$

The input current I_{in} is:

$$I_{in} = \frac{V_{in}}{R}$$

so,

$$V_{out} = -\left(\frac{V_{in}}{R_{in}\times C_f}\right)\times t$$

The graph in Figure 10.13 shows that a constant input produces a voltage ramp output. As we have seen previously, oscilloscopes and XY plotters often use a voltage ramp input such as this so that signals can be displayed with respect to time.

Problem

Consider the integrating operational amplifier circuit shown opposite. The input resistor is 100 kΩ, the feedback capacitor is 10 μF, and the voltage across the input is constant at –2 V.

Calculate the output voltages of the amplifier for each second after the input voltage has been applied, up to 6 seconds. From this draw a graph of output voltage against time and compare this to the input voltage.

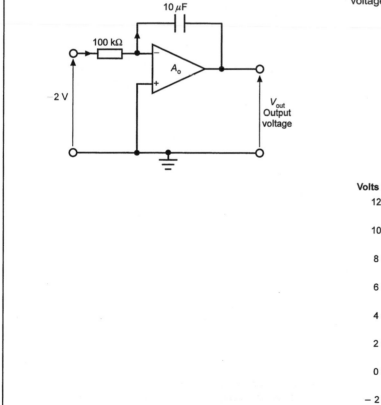

Solution

From the question, we know that V_{in} is a constant –2 V, R_{in} is 100 kΩ, and C_f is 10 μF. To determine V_{out} we use the equation

$$V_{out} = -\left(\frac{V_{in}}{R_{in}\times C_f}\right)\times t$$

The results can be tabulated, and from this the graphs of input voltage and output voltage against time plotted, as shown below:

t (s)	V_{in} (V)	$V_{out}=-\dfrac{V_{in}}{R_{in}\times C_f}\times t$ (V)
0	–2	0
1	–2	2
2	–2	4
3	–2	6
4	–2	8
5	–2	10
6	–2	12

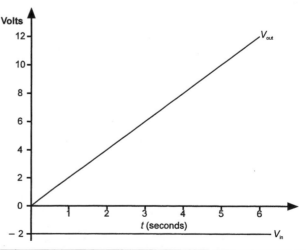

Figure 10.13 Calculation illustrating an integrated output with a constant input voltage

Differentiating amplifier

The differentiating amplifier circuit is the same as the integrating amplifier circuit, except the locations of the capacitor and resistor have been reversed. Figure 10.14 shows a differentiating circuit.

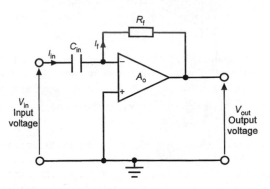

Figure 10.14 Differentiating circuit

The term 'differentiating' refers to the fact that the output voltage is the differential (or rate of change) of the input voltage. That is:

$$V_{out} = -R_f C_{in} \frac{dV_{in}}{dt}$$

Hence if the input voltage remains constant, the output voltage of the circuit tends towards zero over a period of time, t.

Alternatively if the input voltage steadily increases, for example if it is a ramp voltage, the output will stay at a constant value.

To determine the output voltage V_{out} at a given time:

$$V_{out} = I_f \times R_f$$

and

$$I_{in} = -I_f$$

$$I_{in} = \frac{V_{in} \times C_{in}}{t}$$

so,

$$V_{out} = -\left(\frac{V_{in} \times C_{in}}{t}\right) \times R_f$$

Therefore the output voltage of the differentiating operational amplifier circuit is

$$V_{out} = -\frac{V_{in} \times C_{in} \times R_f}{t}$$

The graph in Figure 10.15 shows that a ramp input produces a constant output voltage. Differentiating circuits are often used for control purposes. The circuit shown is, however, very sensitive to noisy signals.

Problem

Consider the differentiating operational amplifier circuit shown opposite. The input resistor is 200 kΩ, and the feedback capacitor is 20 µF. A ramp voltage is applied across the input, which is –2 V after one second. It increases by –2 V every second for six seconds.

Calculate the output voltages of the amplifier for each second after the ramp voltage on the input has been applied. From this draw a graph of output voltage against time and compare this to the input voltage.

Solution

From the question, we know that V_{in} is a ramp increasing by –2 V per second. R_f is 200 kΩ, and C_f is 20 µF. To determine V_{out} we use the equation

$$V_{out} = -\frac{V_{in} \times R_f \times C_{in}}{t}$$

The results can be tabulated, and from this the graphs of input voltage and output voltage against time plotted, as shown below:

t (s)	V_{in} (V)	$V_{out} = -\frac{V_{in} \times R_f \times C_{in}}{t}$ (V)
0	0	–
1	–2	8
2	–4	8
3	–6	8
4	–8	8
5	–10	8
6	–12	8

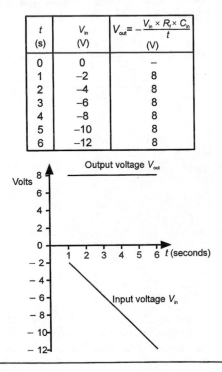

Figure 10.15 Differentiated output with a constant input voltage

Current-to-voltage converter

Some measurement systems produce output signals which vary with the size of the measurand in terms of current. An example of this is the photodiode. The photodiode produces current which is related to the intensity of light falling on it. Figure 10.16 shows a current to voltage converter.

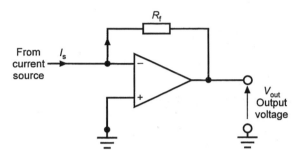

Figure 10.16 Current-to-voltage conversion

All of the supply current passes through the feedback resistor R_f and the inverting input is a virtual earth. Therefore, the output voltage is

$$V_{out} = I_s \times R_f$$

Voltage-to-current converter

The voltage-to-current converter is used when a controlled current supply is needed.

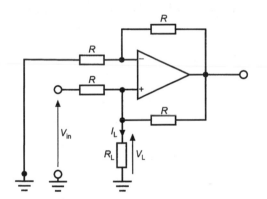

Figure 10.17 Voltage-to-current conversion

Figure 10.17 shows a method of providing a controlled current source, which is independent of load impedance. The load current I_L, is

$$I_L = \frac{V_{in}}{R}$$

Therefore, the load current I_L is not dependent on the load resistance R_L.

Voltage-to-frequency converter

Voltage to frequency converters are often used on the input to digital meter displays. Figure 10.18 shows the operational amplifier circuit which produces voltage to frequency conversion. The voltage-to-frequency converter comprises an integrating amplifier A_1, and a comparator A_2.

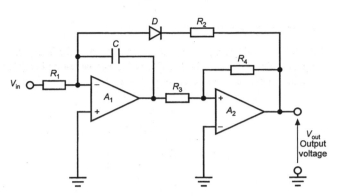

Figure 10.18 Voltage to frequency conversion

Diode D and resistor R_2 provide a discharge path for capacitor C to allow the integrator output to rise to the point where the output of A_2 will switch positive. D will stop conducting, and C will charge from V_{in}. The integrator output will fall at a rate dependent on the time constant CR_1.

When the integrator output reaches the lower switching threshold of the comparator, the output of A_2 will switch negative.

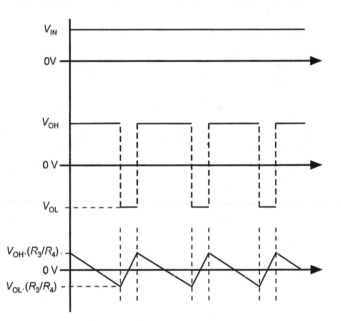

Figure 10.19 Voltage to frequency waveforms

Figure 10.19 shows the waveforms from the voltage to frequency converter. The output frequency is given by

$$f \approx \left(\frac{R_4}{R_3 (V_{OH} - V_{OL}) CR_1} \right)$$

Frequency-to-voltage converter

Figure 10.20 shows how operational amplifiers can be configured as a frequency to voltage converter.

The frequency-to-voltage converter consists of a comparator A_1, and an integrator A_2. The comparator is a zero crossing detector which produces a square-wave output. The output is limited in amplitude (by the back-to-back zener diode pair).

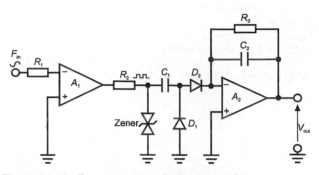

Figure 10.20 Frequency to voltage conversion

When the comparator's output is high, capacitor C_2 is charged through capacitor C_1 and diode D_2. When the output is low, C_1 is discharged through D_1 back to the comparator.

Therefore, C_2 is charged and then discharged through R_3 at a rate which is dependent on the frequency of the square-wave. The average voltage stored by C_2 will increase with frequency, and so V_{out} will be proportional to F_{in}. The output voltage is

$$V_{out} \approx 2 \times F_{in} \times C_1 \times R_3 \times V_{zener}$$

Digital-to-analogue converter

The digital-to-analogue converter is used when a computer or other digital output system is needed to control an analogue device such as a pump or valve.

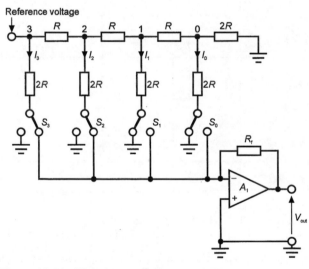

Figure 10.21 Digital to analogue converter

A typical digital to analogue converter configuration is shown in Figure 10.21. It consists of a current-to-voltage converter A_1. The rest of the circuit is in a form known as a ladder network.

The ladder network represents the individual bits of a digital number. Each bit is represented by a current which is fed to the input of the current-to-voltage converter where they are summed together. When a switch is closed it represents logic level 1, and when open it represents logic 0. S_3 is the most significant bit and S_0 the least significant bit. Therefore, in Figure 10.21, the switch configuration currently represents the binary number 1100_2.

Figure 10.24 is an example calculation illustrating digital to analogue conversion.

Analogue-to-digital converter

The analogue-to-digital converter is used when the analogue output from a sensor is to be monitored or displayed by a digital computer, data acquisition system, or other digital input system. Figure 10.22 shows a circuit for a single ramp analogue to digital converter.

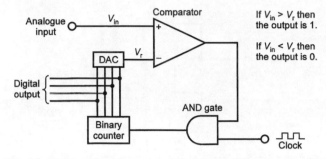

Figure 10.22 Single ramp analogue to digital converter

The analogue signal is converted into a digital number which itself is converted back into an analogue signal. When this analogue signal is the same as the original the process stops, giving an accurate analogue to digital conversion.

When V_{in} is greater than V_r the output of the comparator is logic 1. The AND gate applies a pulse to the binary counter which is synchronous with the external clock. The counter output is incremented by one for each pulse of the clock. The digital output is converted back to an analogue signal which is applied to the comparator until V_r becomes greater than V_{in}. The output of the comparator turns to logic 0 and turns off the AND gate.

The digital output is then equivalent to analogue input V_{in}.

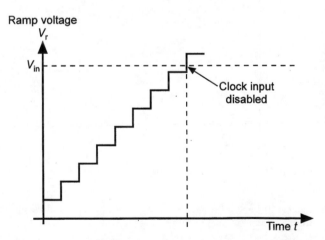

Figure 10.23 Single ramp analogue to digital converter

Problem

Consider the digital to analogue converter shown opposite. Assuming the resistances R are all 1 kΩ, the feedback resistance R_f is also 1 kΩ, and the reference voltage V_{ref} is –16 V, determine the output voltage produced by the digital input 1011_2.

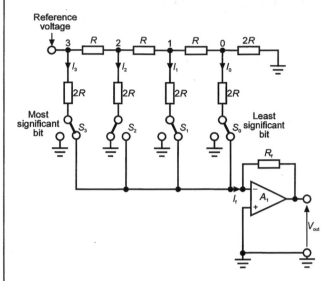

Solution

From the question we know that each of the resistances R = 1 kΩ, R_f = 1 kΩ, and the reference voltage V_{ref} = –16 V.

The current delivered to the summing junction of the current to voltage converter from node 3 of the ladder will be

$$I_3 = \frac{V_{ref}}{2R} = \frac{-16}{2 \times 1 \times 10^3}$$

$$I_3 = -8 \text{ mA}$$

The voltage at node 3 will be the reference voltage V_{ref} which is –16 V. Moving along the ladder, the voltage at nodes 2, 1, and 0 will be half of the voltage of the node to the immediate left. That is:

$$V_2 = \frac{V_3}{2}, V_1 = \frac{V_2}{2}, V_0 = \frac{V_1}{2}$$

Thus

$$V_2 = \frac{-16}{2} = -8 \text{ V},$$

$$V_1 = \frac{-8}{2} - 4 \text{ V},$$

$$V_0 = \frac{-4}{2} = -2 \text{ V}$$

The corresponding currents will be,

$$I_3 = \frac{V_3}{2R}, I_2 = \frac{V_2}{2R}, I_1 = \frac{V_1}{2R}, I_0 = \frac{V_0}{2R}$$

Thus

$$I_3 = = \frac{-16}{2 \times 10^3} = = -8 \text{ mA},$$

$$I_2 = = \frac{-8}{2 \times 10^3} = = -4 \text{ mA},$$

$$I_1 = = \frac{-4}{2 \times 10^3} = = -2 \text{ mA},$$

$$I_0 = \frac{-2}{2 \times 10^3} = -1 \text{ mA}$$

However, because the digital input is 1011_2, the bit corresponding to S_2 is logic 0, so here the current will flow through the 2R resistor direct to earth. Hence the current through R_f will be,

$$I_f = I_3 + I_2 + I_1 + I_0$$

$$= -(8 + 0 + 2 + 1) \times 10^{-3}$$

$$I_f = -11 \text{ mA}$$

The output voltage is

$$V_{out} = -I_f \times R_f = -(-11 \times 10^{-3}) \times 1 \times 10^3$$

$$V_{out} = \textbf{11 volts}$$

Figure 10.24 Digital to analogue converter calculation

Summary

This chapter considered a range of active signal conditioning and interfacing techniques incorporating operational amplifiers. There are many other configurations and techniques available for the large variety of signal conditioning requirements in sensors. There are also other types of amplifier and device which can sometimes be used to achieve the same effect. However, operational amplifier techniques are low cost, readily available and relatively simple.

Questions for further discussion

1. Why do you think integrating and differentiating amplifiers are particularly useful for control applications.
2. It is possible to build similar circuits to the ones we have discussed without using operational amplifiers. Give reasons why you think operational amplifiers are more likely to be used.
3. Consider the differentiating problem given in Figure 10.12. Draw the graph of output voltage against time if a constant input voltage of –2 V is applied, instead of a ramp input. What is happening to the output voltage?
4. Consider the digital to analogue converter example in Figure 10.24. Determine the output voltage produced by the digital input 1100_2.

---

11 Measurement Application Case Studies

In this chapter we will look at three possible situations which require, or could benefit from, some sort of measurement system. We will develop an outline solution for each case. In a practical situation this would provide the basis for a more detailed investigation or design. The design would then be built, tested, modified if necessary, and finally prepared for use.

The type of problems we will look at would normally be solved by a team of designers and engineers. These would each have specialist skills, and by pooling all the available knowledge an effective design should be produced. It is important that a measurement system is designed thoroughly, taking into account all the information available. This saves money and extra work later on, as well as avoiding delays.

Each application will have more than one possible solution. The main purpose of this chapter is to illustrate the initial stages of the design process, and practical application of the topics we have so far covered.

One of the most important stages in applying a measurement system is in deciding exactly what we need to measure. In all systems there will be various parameters, of different magnitudes, scales, and ranges. What needs to be decided early on is the nature of the problem.

When the problem has been defined, there will be various solutions. There will be many factors affecting the choice of final design. It is often not a straightforward or simple choice.

When we have a basic idea of how to solve the measurement problem, the next process is to decide which sensor or sensors to use. As you will have seen in previous chapters, there are many sensors available which detect similar parameters. Here, we rely on knowledge, experience and judgement in determining the most suitable sensor. Again, it may not be a straightforward choice. The sensor which would achieve the best results may be too expensive, or unavailable at the time it is required. Very occasionally, for specialist applications, a new type of sensor may be required.

The choice of display or recording device can be as important as the choice of sensor. A sensed value must be displayed as appropriate to the system. Common factors affecting the choice of display are speed and accuracy.

The signal conditioning function in a measurement system has to match the sensor to the display or recorder. This is often the most complicated part of the design process. Sometimes it will affect the choice of display and sensor. It may need specialist skills, equipment, or information.

After taking into account the sensors, display, and signal conditioning requirements of our proposed solution, we must decide whether it is justifiable. The solution must meet our original objectives. It should be within any limits of cost, performance and efficiency. During the development of the solution, other ideas or possible methods may have arisen.

To develop a solution to each of the three case studies we will try and answer the following questions:

- What is the problem?
- How can we solve it?
- What sensors could we use?
- What type of display is most suitable?
- How do we condition the signals and link the devices?
- Is this the best solution?

Case study: checking the dimensions of a component during manufacture

When a component is being manufactured, it is important to make it to the correct size for its application. This may seem obvious, but apart from ensuring that it fits into where its meant to go, its tolerance also affects the cost of manufacture.

Generally, the smaller the tolerance the more expensive a component is to make.

Quality assurance procedures in manufacturing require engineered parts to meet specified dimensional requirements. Parts are measured and inspected to find whether they are within the tolerance limits of their design.

What is the problem?

A manufacturing plant is producing components in large quantities. Figure 11.1 shows their shape. It also shows the dimension which needs checking at this stage of manufacture.

Currently the measurements of each component are checked by inspectors using instruments such as vernier callipers. In order to make this process quicker and more efficient, a system is needed to show whether or not the dimensions of the component are within tolerance.

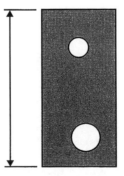

Figure 11.1 Manufactured component

The system must be reliable and need as little maintenance as possible. It should also need little operator skill or involvement, apart from positioning the component, and responding to indications as to whether the component is within tolerance. It is also desirable for the equipment to be made flexible. That is, if the size or shape of component were to change, the system could be readily adapted.

How can we solve it?

One possible solution is to use a displacement sensor. This could be placed above a surface table, forming a 'detector head'. Components can be placed between the detector head and the table. The measurement system will then display to the operator whether the component is within tolerance or out of tolerance. Figure 11.2 shows this arrangement.

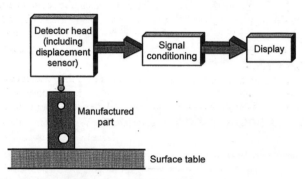

Figure 11.2 Dimension checking

What sensors could we use?

We are measuring a linear displacement, as discussed in Chapter 3. Possible sensors are:

- Dial test indicator
- Linear potentiometer
- LVDT
- Variable area capacitor

First, we need to know how much accuracy is required. Assuming an average tolerance, all the above devices are suitable. There is the cost factor also. We need to keep costs at a minimum but not have these outweighed by regular maintenance costs or poor performance. Since there are likely to be no environmental hazards, maintenance requirements will relate closely to whether or not the sensor has moving parts in contact with other internal components. Of the four, the dial test indicator and the potentiometer have these. The dial test indicator contains a fairly complex mechanical arrangement and the potentiometer has a wiper moving over the resistance element.

For quick and convenient reading it would be appropriate to display the measured value electronically. The dial test indicator cannot do this and reading requires a certain amount of skill from the user. Because of this, and possible maintenance problems, we can discount it.

Each remaining sensor will provide good linearity and repeatability between signal output and displacement. Power supply requirements present no difficulties but they must be stable for reliable results, whichever sensor is chosen. The variable capacitor cannot measure a wide range of displacements. We don't know what the range is, but if the system is to be flexible the small range of the capacitor may limit this. Also electrical

noise can be a problem when signal conditioning. Thus the variable area capacitor is not the most suitable choice.

In terms of cost, the LVDT will be the most expensive and the potentiometer the least expensive. However precision is a key requirement and the LVDT is the best of the four in this regard. It also has a higher resolution and requires less maintenance than the potentiometer.

After looking at the main factors affecting the choice of sensor, the LVDT is the most suited to this application.

What type of display is most suitable?

The LVDT output is electrical. With suitable choice of interfacing and conditioning circuitry we can use the sensor signal to drive an electronic display. This could indicate the actual displacement of the surface of the component to the detector. However, the operator then has to decide whether the component is within tolerance and possibly perform a calculation. It would be desirable for the measurement system to do this and save operator time. Hence a simple means of indicating whether the component is within tolerance or not would be desirable.

In Chapter 8 we saw that the liquid crystal display provides precise indication with low power requirements. An LED display could also be used. Even though it needs more power than a liquid crystal display it has a quicker response time and is more visible. Both LCD or LED displays could be used here since we could arrange segments to have the words 'pass', 'oversize' or 'undersize' to appear in the readout when appropriate. If the components pass they can continue to the next stage of manufacture. If they are oversize they can be reworked as appropriate. If they are undersize they may be of use elsewhere or need to be scrapped.

A much less expensive means of indication is to simply switch on a green light when the component is within tolerance, an orange light when it is too big, and a red light when it is too small.

To keep costs down, initially we could just use two lights. A green light would indicate whether the product is within tolerance and a red light would indicate whether it is too big or too small. This is the technique we will investigate.

How do we condition the signals and link the devices?

Appropriate signal conditioning may be built into the LVDT system when it is purchased. If not, it may be possible to buy a circuit with the LVDT or the display which conditions the signals appropriately. If not, an electrical or electronics engineer would need to design an appropriate circuit. We will assume this is the case.

The position of the detector head can be accurately set and calibrated so that the dimension to be measured gives an LVDT zero output when it is correct. Oversize will produce a positive output and undersize a negative output.

From here on the signal conditioning requirement of the system starts to get complicated. You may be unfamiliar with some of the terminology and devices used. Remember, in a practical situation an electrical or electronics engineer would probably design the signal conditioning circuit, so you may not be required to fully understand it.

The specification for our system requires the display to indicate that the dimension is within tolerance or out of tolerance. We can detect these conditions and switch the appropriate signal to the display by means of a window comparator, as shown in Figure 11.3.

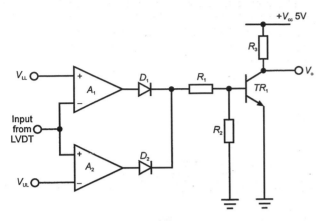

Figure 11.3 Window comparator

The reference voltage V_{LL} is equal to the LVDT output at the lower tolerance limit and V_{UL} equal to the LVDT output at the upper tolerance limit. Hence we can use the LVDT to drive the window comparator directly. This is by earthing one of its outputs and connecting the other to the comparator signal input. If the LVDT output is less than the lower limit, the output of comparator A_1 will be high, forward biasing diode D_1 and switching on transistor TR_1. V_o will be low. If the LVDT output is greater than the upper limit, comparator A_2 output will be high, D_2 forward biased and V_o low. Where the LVDT output is between V_{LL} or V_{UL} both comparators will be output low, TR_1 will be off and V_o will be high.

Outside of the tolerance band one comparator will be output low and the other output high. In these cases diodes D_1 and D_2 prevent the high output comparator sending current to the low output comparator. Hence they allow the comparator outputs to be logically connected.

The green and red lights can be driven from V_o as shown in Figure 11.4.

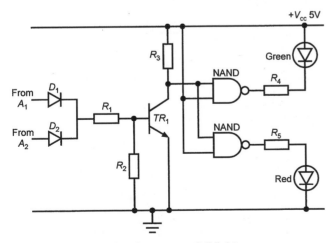

Figure 11.4 Switching for pass or fail lights

TR_1 is off when the LVDT output is between V_{LL} and V_{UL}. Both NAND gates have one input connector to the positive rail (logic '1') and both are connected to the collector of TR_1. Thus when TR_1 is off the NAND gates will have both inputs high and therefore outputs low.

In this condition the green light will come on and the red light will be off. When TR_1 is on both NAND gates will have one input at logic '0' and their outputs will go high. The green light will be off and the red light will be on.

Is this the best solution?

The solution is an improvement over using a vernier calliper to check the dimensions of every component. It is quicker, less prone to error, and needs less operator skill. However, it can only measure one dimension of the component whereas an operator using a vernier calliper could measure several in succession. To enable the suggested solution to do this would need several different detector heads. Alternatively one detector head could be made flexible so it can determine whether more than one dimension is within tolerance, but this would require more complex signal conditioning circuitry.

The solution does not indicate whether the component is oversize or undersize. This would need to be found out by a later process. If this was to be performed by, say, an operator using vernier callipers, then the proposed solution is an improvement but not ideal.

Determining whether the component is oversize, undersize, and perhaps indicating by how much, would be a more suitable solution. It should also be able to adapt to read more than one component dimension. The setting up costs and maintenance would be more expensive, but this would be outweighed by increases in efficiency and speed.

This does not appear an ideal solution. There seems room for improvement and so further investigation into this problem would be worthwhile.

Case study: vibration monitoring on a road bridge

The performance and reliability of structures such as road bridges, tunnels or high-rise buildings depends very much upon how the design withstands the vibrations set up in the structure. These vibrations will occur in normal use for a number of reasons. Vibrations occur in a building mainly because of windy weather but also other factors such as passing traffic or underground transport systems. They occur in tunnels because of traffic movement within the tunnel, and to a lesser extent above it. Vibrations occur in bridges because of weather conditions, traffic on the bridge, and other factors such as waves hitting its supports.

In the past there have been some spectacular, and often disastrous bridge collapses, sometimes causing loss of life. In several cases this was because the design of the bridge could not withstand the effects of the vibrations it was subjected to.

What is the problem?

A bridge is made up of a number of members, assembled into frameworks. Some members act as ties. These hold parts of the bridge together. Other members act as struts. These hold parts of the structure apart. Other members act as horizontal beams or vertical load-bearing supports.

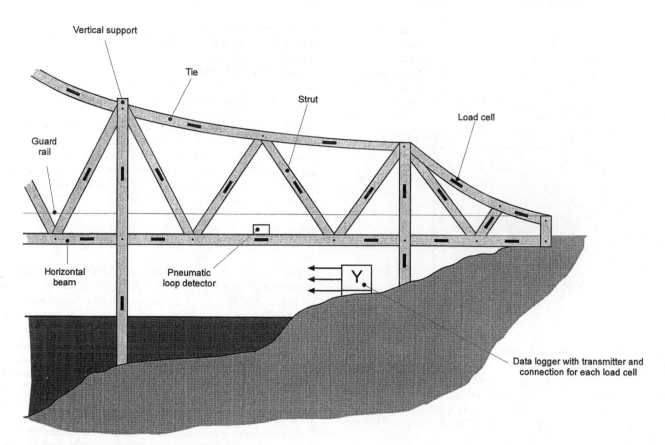

Figure 11.5 Bridge structure monitoring

Ties are always in tension, that is, the forces acting upon them tend to increase their length. The forces in struts are compressive and tend to shorten them. In normal operation the movement of traffic causes vibrations in the structure. This results in quickly changing forces in the members of the bridge in addition to the loads they already bear.

We need to design a measurement system to carry out a short-term test on a road bridge structure. It needs to monitor vibrations so researchers can find out what sort of traffic, and in what quantities, causes the most vibrations. It should also show the worst type of vibration the bridge is subjected to.

How can we solve it?

Bridges are largely made up of a number of similar frameworks. Because of this we only need to monitor the performance of one section. Other sections are assumed to behave in a similar manner.

To monitor vibration thoroughly, we need to sense it in each member of the section we are monitoring. We can do this by using a group of sensors to measure the changes in load of each of the section members, over a given time period.

When we are measuring the vibrations in the bridge, we can also measure the traffic passing over it. We can detect and count the number of different categories of vehicle which pass over the bridge section. Each category of vehicle will be within a defined range of weights. For example, cars may be defined as weighing under seven tonnes, small trucks seven to fifteen tonnes, medium trucks fifteen to thirty tonnes, and so on.

By measuring vibration in the bridge while at the same time measuring the traffic passing over it, we should be able to see and understand what sort of vehicle produces what sort of vibration.

Hence we can see which vehicles produce the largest vibrations, and the largest vibration the bridge is subjected to over the given time period.

What sensors could we use?

Load cells and accelerometers can sense vibration. Both provide the necessary precision and accuracy and require little maintenance. The load cell, however, has a cost advantage and so we shall use it to detect the vibrations in this application. These would be positioned on the members of the bridge as shown in Figure 11.5.

The weight of each vehicle passing over the bridge can be detected by placing a pressure pad on the road surface. Figure 11.6 shows the form of this pressure pad, which stretches across the carriageway. This type of sensor is known as a pneumatic loop.

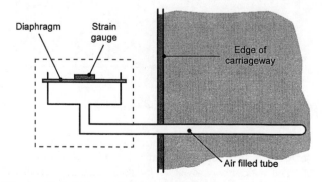

Figure 11.6 Pneumatic loop sensor

When a vehicle passes over a pneumatic loop sensor, it applies a force to the air-filled tube. This changes the pressure on the diaphragm causing it to flex. A strain gauge on the diaphragm momentarily changes its resistance in proportion to the force applied. The strain gauge forms one arm of a Wheatstone bridge. An output is produced as each vehicle axle passes over the sensor. The size of the output is proportional to the weight of the axle, and hence gives an indication of vehicle size.

What type of display is most suitable?

Most cars will produce two outputs when they drive over the pneumatic loop sensor, because they have two axles. Trucks usually have three or more axles, so will produce more outputs. In order to ensure the system does not incorrectly record too many trucks, the display must in some way indicate the number of axles of similar weight passing close together.

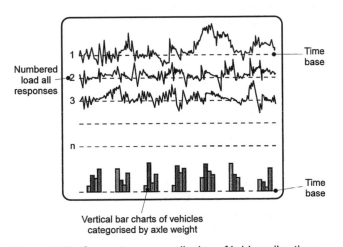

Figure 11.7 Computer screen display of bridge vibrations and traffic

This application does not require a real-time display, although one could be provided. It will be more appropriate, since the test will be carried out over a matter of hours, to use a data logger to record the data. It can then be analysed later in a laboratory.

Because there is a large amount of data being produced through a number of different channels, a computer based data acquisition system would be the best option. It would be possible to use analogue meters and chart recorders. However, a computer based system allows us to readily manipulate the data and display it in several different ways.

A parallel representation of all of the channels would be needed. This would be provided on a computer screen, as a group of varying signals against a time base. The waveforms would scroll to the left as time progresses.

Part of the screen can be devoted to a bar chart display. This would display the weights of the axles passing over the pneumatic loop sensor. If three or four similar, heavy weights appear close together, it would be reasonable to assume they are the same vehicle. This presentation would be regularly refreshed. The display would be as shown in Figure 11.7.

How do we condition the signals and link the devices?

The electrical outputs from the load cells and the pneumatic loop sensor need to be converted to digital data. This will then be suitable for input to the data logger. The first action will be to amplify the load cell bridge outputs. This can be done using an instrumentation amplifier.

The amplifier outputs will be sampled by a multiplexer, fed in turn into a sample-and-hold circuit and then converted to digital data by an analogue-to-digital converter. The multiplexer, sample-and-hold and analogue-to-digital converter will all reside on a data acquisition board fitted into the data logger. The process for vibration recording is as shown in Figure 11.8.

The size, frequency, and range of vibrations produced by the traffic travelling over the bridge will vary considerably. All the signal conditioning equipment must be fast enough and capable of dealing with these signals.

In order to count the numbers of the different types of vehicles, the pneumatic loop output pulses will be converted to digital values, the magnitude of which will indicate axle weight. The computer software programme will check the value and increment the appropriate counter, one of which will be provided for each vehicle weight category. At each display refresh point the counter value will determine the height of the bar graph for that vehicle category. The counters are reset to zero after each display refresh. The process is shown in Figure 11.9.

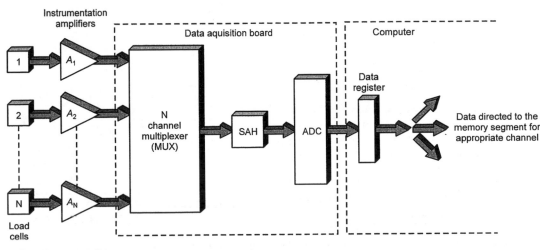

Figure 11.8 Vibration recording

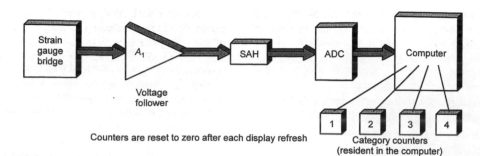

Counters are reset to zero after each display refresh

Category counters
(resident in the computer)

Figure 11.9 Determining traffic count

Is this the best solution?

Using a computer based data acquisition system is a convenient and versatile method of collecting the data required. It allows detailed analysis of the vibration caused by traffic, showing what type of traffic causes what type of vibration. The volume and detail of information required is high, and the techniques discussed are a cost effective and efficient solution to the problem.

Case study: wheel speed measurement in an anti-lock braking system

Anti-lock braking systems (ABS) are now included in many types of car. They prevent the wheel of a car locking, and so skidding when the brakes are applied. If a car goes into a skid it is difficult to control and overall stopping distance is increased. Hence anti-lock braking systems on cars are a useful safety feature which may have prevented, or lessened the seriousness of, many accidents.

What is the problem?

A sensor is required to measure the speed of rotation of motor car wheels. A control module is already available. It requires a pulsed input signal with two levels; 0 volts and 5 volts. It will then compare the sensor signals from each wheel and lift the brake on any wheel which has locked, or varies significantly from the others in rotational speed, over 15%. This action is designed to prevent skidding and thus enhance operational safety.

The sensor chosen must be able to operate over a wide temperature range. It could be used in countries where the temperature drops to say, 240 K (about −33 °C), or other countries where the temperature, not including any heat generated by the engine, reaches, say 313 K (about 40 °C). It must not be affected by vibration generated from the road surface or engine. It must also function in an environment which can be dirty, dusty, and be subjected to snow and ice, oil, and salt water.

Because this is a safety device, high reliability and low maintenance are essential.

How can we solve it?

A sensor that produces an electrical output proportional to wheel speed can be fitted on each wheel axle. All four signals can be conditioned and fed into a controlling module for comparison. Outputs from this module will control the wheel brakes individually.

If a wheel sensor produces a signal indicating that one wheel has stopped or is rotating at a significantly different speed from the other three wheels, it identifies a potential skid. The control module will detect this and release the wheel brake accordingly, until corrected.

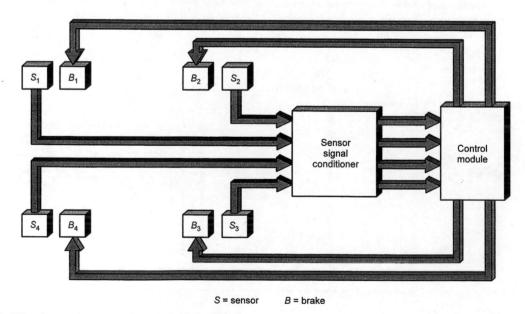

S = sensor   *B* = brake

Figure 11.10 Wheel speed sensing for anti-lock braking system

Figure 11.10 shows the system we shall use. S_1, S_2, S_3, and S_4 represent the sensors on each wheel. B_1, B_2, B_3 and B_4 represent the brake on each wheel.

What sensors could we use?

We are sensing angular velocity. The sensors we could use are therefore angular displacement sensors or proximity sensors. Possibilities include:

- D.C. tachometric generator.
- Optical shaft encoder with LEDs and light detectors.
- Magnetic reed switch sensor.
- Hall-effect proximity sensor.
- Variable reluctance proximity sensor.

The output of a d.c. tachometric generator is an electrical signal whose voltage is proportional to rotational speed. Its polarity is related to the direction of rotation. The output voltages from tachometric generators mounted on each wheel could be readily compared in the ABS system in order to detect a wheel stopped or travelling at a different speed.

It would require an arrangement such as a toothed wheel on the drive shaft, to engage a pinion on the tachometric generator shaft. Tachometric generator shaft speed may need to be a greater than the vehicle wheel speed, so gearing might be required between them.

Certain parts of the d.c. tachometric generator will be subject to wear. Regular maintenance and replacement would be required. It also needs to be sealed to prevent dust entering as this increases the wear rate on moving parts. A further disadvantage is that the copper and iron used in its construction mean that it is relatively heavy.

Signal conditioning should be fairly straightforward. However, the regular maintenance required, and possible failure of the device if this is not performed, discounts the d.c. tachometric generator.

Another possibility is the optical shaft encoder. The disk would be mounted on the hub of each wheel of the car. An incremental encoder disk could generate a pattern of 16 pulses for every revolution of the disk. The LEDs and the light detectors could be mounted either side of the disk on the axle support.

The relationship between the slotted disk speed and the digital code generated is linear. If the Gray code output is converted to binary, the possibility of errors in counting the position code would be minimised. Because there is no physical contact between the LEDs, the light detectors and the slotted disk, very little mechanical wear will occur. Hence maintenance will be low. However, the 4-bit encoder assembly would have to be completely sealed to exclude dust and moisture.

The performance of LEDs and light detectors (such as phototransistors) is affected by temperature changes. Vibrations can shorten their lives, and may also seriously affect the operation of the system. To overcome these problems would be expensive, so this solution is unsuitable.

A magnet could be embedded or attached to each wheel hub or axle, and a magnetic reed switch mounted nearby on each axle support. Every time the magnet passes the reed switch, the contacts would momentarily close, temporarily making a circuit. A pulsed output is then produced as each wheel turns.

Magnetic reed switched have a fast response and are inexpensive. They are non-contact so would not be subject to wear. They are also reliable and not affected by dirt or dust. Unfortunately, they are fragile and can be damaged by vibration. They are likely to break if hit by an object such as a stone thrown up from the wheel. Also, because the opening and closing of the reed switch is mechanical and this application requires a high rate of switching, life expectancy may not be as long as some of the other devices. This reduces their reliability and so reed switches would not be suitable for this application.

The arrangement of variable reluctance and Hall-effect proximity sensors would be similar. They would both consist of a probe assembly mounted on the axle support of each wheel. They would be set to detect the motion of a toothed wheel mounted on each wheel drive shaft. The toothed wheel would be manufactured from a suitable ferrous material such as steel. A pulse would be produced each time a wheel tooth passes the sensor. If the wheel has, for example, 50 teeth, then 50 pulses would be generated for every revolution of the car wheel.

Each wheel will generate a frequency signal and when any wheel has a frequency that is different from the others, the control system will operate. The relationship between toothed wheel speed and the frequency of the signal generated would be linear. There would be no contact between the probe and the toothed wheel and hence no mechanical wear.

Both variable reluctance and hall effect devices are reliable and require little maintenance. The probe assemblies of both types of device would need to be reasonably sealed to exclude dust and moisture. They would not be affected by temperature changes or vibration.

Hall-effect devices have better signal to noise ratios than variable reluctance probes and are suitable for low speed operation. However, variable reluctance probes are less expensive, more robust and simpler. By using a high number of teeth, they would adequately detect the rotation of each wheel even at low speeds.

Of the devices we have considered to detect the rotational velocity of the car wheels, the Hall-effect or variable reluctance proximity sensors seem the most suitable. For reasons of simplicity and cost, we shall use variable reluctance probes.

What type of display is most suitable?

In this case no display is required.

How do we condition the signals and link the devices?

The outputs from the variable reluctance probes will be a series of pulses. Each pulse is generated as a tooth approaches and passes the pick-up coil. Figure 11.11 shows how the variable reluctance probe could be interfaced to a control module.

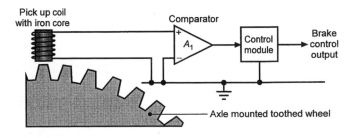

Figure 11.11 Linking the variable reluctance probe to the control module

If we looked at the shape of the pulse produced on an oscilloscope, we would find that they will tend to be of a triangular or sinusoidal shape. However, it is the frequency of the signal that is required and we can use a sine-to-square wave comparator to provide trains of output pulses. Each pulse would have 0 and 5 volt levels. This would satisfy the input requirements of the control module.

Is this the best solution?

The variable reluctance sensor would send suitable signals to the control module and hence effectively form part of an ABS system. It is reliable, requiring little maintenance, and is relatively inexpensive. However, because this is a safety related problem where the driver of the car would get used to the response of the vehicle with the ABS system, effective and reliable performance should have a higher priority than price alone.

This solution initially seems the best, but because it is safety related, a more thorough investigation would usually take place to confirm this.

Summary

In this chapter we have looked at three case studies of measurement applications. Using a methodical approach we have discussed possible sensors which could be used in each case, and outlined possible solutions. After discussing these solutions we have commented on their suitability and decided whether further investigation is needed before producing a detailed design.

There are a number of factors which influence the choice of sensor and devices within measurement systems. Particularly with sensors, there is often no obvious solution. This makes it important to weigh up the benefits and possible drawbacks of each device in turn.

By using a methodical approach a suitable choice can be effectively and efficiently made. Sometimes a solution which initially seems obvious may, after investigation, not seem so suitable. For the best choice to be made as much information about the problem and all possible solutions should be taken into account.

Questions for further discussion

1. The approach to these case studies was to attempt to answer a number of set questions. Are there any further questions which could be added to produce more effective solutions?
2. Do you think there are any major faults with the choice of display in the first two case studies?
3. Consider the case study concerning the size of a manufactured component. What advantages could there be in using a dial test indicator instead of the other sensors discussed?
4. Accelerometers are sometimes specifically designed to measure vibration. Discuss the changes required if accelerometers were used instead of load cells in the case study measuring vibrations on a road bridge.
5. In the anti-lock braking system case study, why do you think an a.c. tachometric generator was not discussed as a possible choice of sensor?
6. In the road bridge case study, what sensors could be used simply to count the traffic passing over the bridge? Methodically choose a suitable sensor and outline a possible solution.

12 Control Application Case Studies

This chapter looks at case studies in which sensors form part of control systems. In a similar manner to the measurement applications in Chapter 11, we will develop an outline solution for each case.

Dividing applications into either measurement or control is somewhat artificial, since the sensor does the same job. Whether the application is a pure measurement system, an open-loop control system or a closed-loop control system, the sensor detects a signal or stimulus and from this produces a measurable output. Beyond this point, however, the control system is dedicated to controlling the physical variable, not simply recording and displaying its value. For this reason, signal conditioning and interfacing will be a little different for a control system. This is the main difference when choosing sensors for control rather than measurement.

In an open-loop system, such as the street lamp example we looked at in Chapter 2, the output value of the system does not need to be measured. Hence in many open-loop systems sensors are not necessary. Sensors used in open-loop systems, will only measure the input signal, if at all, or purely display the output value as in a measurement system. If sensors feed back a value for control purposes, the system will then be under closed-loop control.

Although not usually necessary it is often useful to display or record data from sensors used in control systems. Computers are commonly used in control systems because they can quickly process and respond to large amounts of data received from sensors.

We shall use the same methodical approach to choose sensors for control applications as for measurement applications. That is:

- What is the problem?
- How can we solve it?
- What sensors could we use?
- What type of display is most suitable?
- How do we condition the signals and link the devices?
- Is this the best solution?

In this chapter we will look at sensors used in one open-loop and two closed-loop control systems.

Case study: inspection on a production line

Many production lines are largely automated. However, visual inspection by a human operator is often necessary during certain stages of manufacture. This is to check for say, scratches on a component, chips in paintwork, or missing parts of an assembly. These can be quickly checked by the human eye but would be difficult or time-consuming to automate.

In factories producing large numbers of similar components, conveyor-belts are often used. These not only transfer components from one operation to the next, they also allow quick and simple operations to be performed efficiently with minimum disruption to production.

What is the problem?

A factory produces medium-sized batches of three different types of product: product A, product B, and product C. It only produces one type of product on each batch run. Production is mainly automated. However, at one stage the products have to be visually inspected by a human operator. The previous process automatically places the individual products on a short conveyor-belt at 5-second intervals. The system must allow adequate time for inspection of each product, but not be too slow. The conveyor-belt is driven by a d.c. motor.

If product A is being produced, visual inspection only takes a few seconds. Visual inspection of product B takes up to a minute, while visual inspection of product C can take several minutes. A control system is needed which allows the human inspector adequate time to visually check each product.

The speed of the conveyor-belt needs to be displayed locally. This is for use with time and motion studies, to help predict the time scale of the current job, and to allow inspection rates to be compared.

How can we solve it?

A simple open-loop system could be used to control the speed of the conveyor-belt. A basic system would have three set-speeds, one for each type of product. However, because operators work at different rates, and product design may change slightly between batches altering inspection time, a variable speed control could be used.

The conveyor-belt would be set to run quickly for batches of product A, less quickly for product B, and slow speed for product C to allow adequate time for inspection. This would simply be a speed control on the conveyor-belt motor, set by the operator. An emergency stop button would also be necessary. This is to reduce the chance of injury in case of an accident involving the machinery. Most motors will already have some form of speed control and emergency stop, if not it would be a straightforward task to add one. The motor speed would be controlled, for example, by adding a potentiometer to adjust the motor supply voltage.

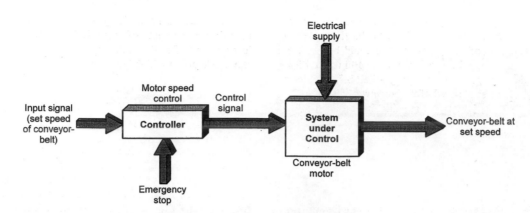

Figure 12.1 Open-loop control system to control the speed of a conveyor-belt

Even though an open-loop system is to be used, it is still necessary to sense and display the speed of the conveyor-belt locally. Figure 12.1 shows a block diagram of the proposed open-loop system. Figure 12.2 shows the proposed conveyor-belt system.

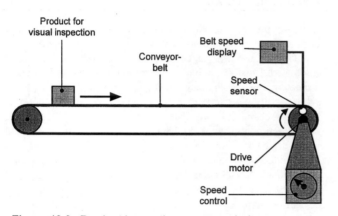

Figure 12.2 Product inspection conveyor-belt

Closed-loop control of the system is not necessary because an operator is always monitoring the situation. Hence the consequences of the system going out of control would not be serious because the operator would take immediate corrective action.

What sensors could we use?

Relying on the setting of the motor speed control alone would not give an accurate indication of conveyor-belt speed. This is because it would vary as products of different weights on the conveyor-belt vary the load on the motor. The conveyor-belt would slow down or speed up with different products but the speed control would be in the same position. Hence a sensor is required.

To detect the speed of the conveyor-belt we could use a sensor on the drive shaft of the motor, or on the conveyor-belt itself. The accuracy, precision and sensitivity do not have to be high, and it is important to keep costs low. Possible sensors we could use are:

- A.C. tachometric generator.
- D.C. tachometric generator.
- Optical shaft encoder with LEDs and light detectors.
- Magnetic reed switch sensor.
- Hall effect proximity sensor.
- Variable reluctance proximity sensor.

If we are to use a tachometric generator, it would link directly to the motor drive shaft, or to the conveyor-belt itself by means of a contacting wheel or similar. It would then produce an electrical signal which relates to speed.

A.C. tachometric generators are generally cheaper than d.c. tachometric generators which is a major factor affecting the choice in this application. The output from an a.c tachometric generator is also less electrically noisy. A d.c tachometric generator could give an indication of the direction of the conveyor-belt but this is irrelevant in this application. We can dismiss the d.c tachometric generator because of its relatively high cost.

Optical encoders could be used by incorporating a coded disk on the drive shaft of the motor, or by encoding one edge of the conveyor-belt. They would provide a high degree of accuracy, but this is unnecessary in this application. The costs of the encoder and signal conditioning make this option too expensive to use here.

Magnetic reed switch sensors are low cost, simple devices. In this application a magnet or magnets could be embedded in the motor drive shaft or at set distances along the conveyor-belt. This would produce a pulsed output proportional to the speed of the belt. Reed switches have a relatively long working life compared to other types of switch. However, they still have a mechanical movement of the reeds which reduces their life expectancy compared with other proximity sensors. They are also fragile and would need protection from knocks or vibration inherent in this application. Therefore they can be discounted.

Both Hall-effect and variable reluctance proximity sensors could monitor the rotation of a toothed wheel attached to the conveyor-belt drive shaft. This would be in a similar manner to the ABS rotation sensing technique we discussed in Chapter 11. However, the rotational speed would be much slower in this application, especially when product C was under inspection. The variable reluctance sensor does not function well at low speeds. Hall-effect proximity sensors could function well here but are relatively expensive.

The simplest and cheapest method of monitoring the speed of the conveyor-belt is to use an a.c. tachometric generator attached directly or by a gear wheel to the conveyor-belt motor drive shaft.

What type of display is most suitable?

There are several types of display we could use here but the emphasis must be on simplicity and cost. The display needs to be mounted where it can be easily seen by the operator. It should also be able to be seen by say, an engineer, so they can asses how long each part takes to inspect without interrupting the operator.

The units used to display conveyor-belt speed do not have to be in standard units. Because the speed relates to the time taken for product inspection, an arbitrary linear scale could be used. For example, this could simply be a scale from 0 to 10 in steps of say, 1. Here, 0 would correspond to the conveyor-belt being stationary and 10 would be the fastest speed of the conveyor-belt.

To keep costs low and the signal conditioning requirement at a minimum, an analogue display is desirable. Possible analogue displays we could use are:

- Moving coil meter.
- Moving iron meter.
- Oscilloscope.

The moving coil meter could display the amplitude of the output signal from the a.c. tachometric generator. This would be proportional to the speed of the drive motor shaft and hence the conveyor-belt. The display scale would be calibrated in terms of conveyor-belt speed. These meters are inexpensive and are available in a number of sizes and ranges.

The moving iron meter would function in a similar manner to the moving coil meter. However, they are more suited to non-linear scales, and the moving coil display is more suitable.

The oscilloscope would display the output waveform from the a.c tachometric generator, as the amplitude and frequency of this waveform relates to the speed of the drive motor shaft and hence the conveyor-belt. The oscilloscope screen could be calibrated with respect to speed. This option is not as easy to read as the other analogue meters and is also more expensive. Hence the oscilloscope is not the best choice of display.

For this application a moving coil meter with a simple linear scale provides an inexpensive but effective solution to the display requirements.

How do we condition the signals and link the devices?

Moving coil meters can usually only accept d.c. input. Hence the output from the a.c tachometric generator will have to be converted to d.c. To do this we can use a device called a rectifier. This is a low cost, simple device which effectively changes a.c to d.c.

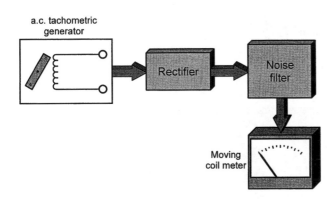

Figure 12.3 Conveyor belt speed sensing signal conditioning

The a.c tachometer output waveform will be proportional to speed in terms of amplitude and frequency. The most simple technique is to use the amplitude and display this directly on the moving coil meter. Figure 12.3 shows the basic signal conditioning of this application.

Because a.c. tachometric generators sometimes produce an undesirable amount of electrical noise, a noise filter is incorporated after the rectifier. The type of moving coil meter chosen should be a high impedance load.

Is this the best solution?

A closed-loop control system could be used in this application but it would be expensive compared to the open-loop solution discussed. The application did not demand the level of control provided by a closed-loop system, and cost also had to be low. The operator is the best person to control the speed of the conveyor-belt, as inspection times vary between products and batches. Hence the open-loop option was taken

The solution has produced an open-loop control system incorporating a measurement system. The measurement system gives the required level of accuracy and costs have been kept to a minimum. Hence this is a simple and effective solution.

Case study: control of thickness in sheet steel manufacture

One of the stages in the manufacture of sheet steel involves passing a strip of hot metal between two rollers. This is to reduce its thickness and improve its mechanical properties. To produce a high quality product this steel should be within a high tolerance. Hence accurate and precise control of the rollers is necessary.

In a steel mill the metal would travel through a number of sets of rollers until it was at the required thickness. In this example we will concentrate on a single set of rollers.

What is the problem?

Figure 12.4 shows the basic configuration of steel mill press rollers. In steel strip manufacture the thickness of the strip depends upon the press roller spacing. Control of the thickness of the strip is achieved by setting the position of the moveable press roller relative to the fixed press roller.

A control system is required which is able to detect variations in the thickness of the steel strip after it leaves the press rollers. It must then regulate the moveable press roller position, to keep the thickness at the required value

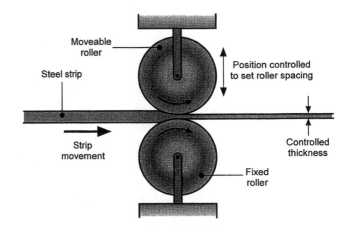

Figure 12.4 Setting steel strip thickness

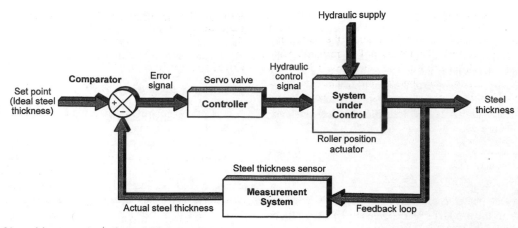

Figure 12.5 Closed-loop control of steel thickness

How can we solve it?

A closed-loop control system will provide a solution to this problem. Figure 12.5 shows the block diagram. Variation in steel thickness will be sensed and converted into an error signal by comparing the actual thickness of the steel to the required thickness (the set point). After suitable signal conditioning and interfacing the control signal then controls the roller position actuator.

Key fact

An actuator is a device in a system or process which has an affect on the parameter under control. In a simple closed-loop control system an actuator usually produces an output, such as mechanical movement, from the control signal.

In this case the roller position actuator lifts or lowers the moveable roller. Hence it produces a mechanical movement from the control signal to alter the roller position appropriately. This will ensure the thickness of the steel is of the value required.

In this working environment the rollers are moved by a hydraulic power source. A hydraulic servo valve will be needed to interface between the electronic circuitry and the actuator.

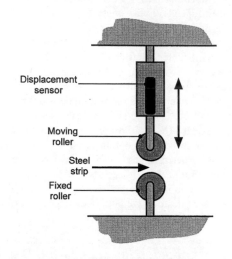

Figure 12.6 Steel thickness sensor assembly

Once the steel has been through the rollers, it passes between two smaller rollers or followers. The purpose of these is to convert variations in the thickness of the steel into linear displacements. A linear displacement sensor is then required to measure these thickness variations. Figure 12.6 shows this steel thickness sensor assembly. Figure 12.7 shows the complete solution outline.

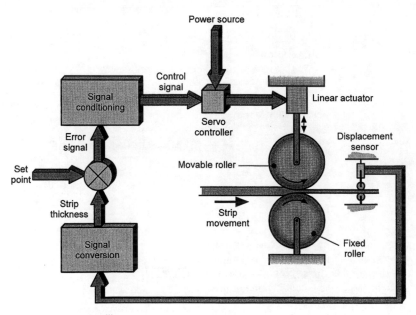

Figure 12.7 Thickness control in a steel mill

What sensors could we use?

A linear displacement sensor is required. Since the steel strip moves at high-speed through the rollers a fast response with negligible hysteresis will be needed. The environment is harsh; having varying degrees of hot, humid, dirty, and corrosive conditions. The sensor must be able to withstand these conditions with high level of reliability. The sensor must have a high degree of accuracy and resolution. The variation in steel thickness should not be large so the sensor only has to operate over a small range.

Because the sensors are operating in a harsh environment, they are not likely to have a long operating life. Therefore, their cost would ideally be low as they will often need to be changed. However, low cost should not be at the expense of performance; incorrect results could results in a great deal of expense. Some form of protection and temperature stabilising around the sensor would be desirable to improve repeatability.

Possible linear displacement sensors we could use are:

- Linear potentiometer.
- LVDT.
- Variable area capacitor.

Linear potentiometers generally have a cost advantage over the other types of displacement sensor. However, in this application, high specification linear potentiometers would be required to provide the necessary accuracy and resolution. Certain types of potentiometer, such as those having plastic components, may not withstand the heat of the steel rolling process. The linear potentiometer has a wiper moving over the resistance element which is therefore subject to wear. For these reasons the linear potentiometer is not the best choice.

An LVDT would produce the necessary accuracy, precision, and repeatability required. It would also cope with the harsh environment better than the potentiometer, but would still need some protection. LVDTs are expensive however, and because they will still need regular replacement or special protection at extra cost, they do not seem the ideal choice.

The variable area capacitor is a non-contact device particularly suited to harsh environments. It can withstand high temperatures, vibration and humidity. The cost of protecting an LVDT to endure this environment makes the variable capacitor the less expensive choice. Only a small displacement needs to be measured which is within the capacitors range. They also have infinite resolution and are sensitive, giving a fast response. Hence the variable area capacitor is the most suitable displacement sensor for this application.

What type of display is most suitable?

The basic control process does not need a display. However, so the steel can be checked for quality and the process monitored for potential problems such as wear in the rollers, periodic recording is desirable.

Information from the displacement sensor could be collected on a selection of the strips rolled on any given day. This could be achieved by moving coil, servo chart, ultraviolet or thermal array recorders, an XY plotter or a data acquisition system.

The changes in linear displacement would be fast. The response time of the moving coil meter is too slow to record the variations in steel thickness adequately.

The application does not justify the expense of an ultraviolet recorder. There may also be problems associated with storing the light sensitive paper in a steel mill.

XY plotters are relatively slow and need regular maintenance.

The servo chart recorder and the data acquisition system are all suitable for monitoring the process. The decision is based on cost, if a hard copy only is required, or the data is to be retrieved for later use. If only hard-copy is required the servo chart recorder would be adequate. For storage and retrieval of data a high specification thermal array recorder or a data acquisition system would be necessary.

The servo chart recorder is low cost, portable and easy to use. It can adequately provide a hard copy in real-time. Assuming no data retrieval or storage is necessary, this would be a suitable recording device for this application.

How do we condition the signals and link the devices?

The capacitive sensor will be connected as one arm of an a.c. bridge circuit. This is a circuit similar in principle to the Wheatstone bridge, but in this case instead of finding out the value of an unknown resistance, it determines the value of an unknown capacitance. When the strip thickness is at the required value the output from the bridge circuit will be zero and it will be balanced.

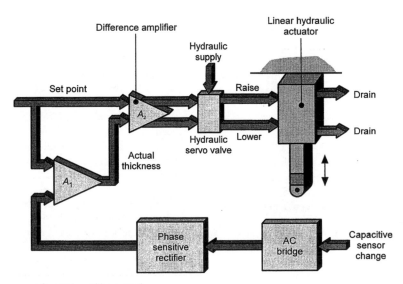

Figure 12.8 Capacitive sensor for strip mill control

If the strip thickness exceeds or is less than the set value the bridge will produce an output, the phase of which will depend on whether the thickness is above or below the set point. This determines the direction in which we should move the roller, to restore the desired thickness value.

A phase-sensitive rectifier is a device which will produce a d.c. output from an a.c. input, the polarity of which depends on the a.c. phase. Connecting the output of the a.c. bridge circuit to a phase sensitive rectifier provides a d.c. signal proportional to the change in capacitance, the polarity of which will depend on the direction of the capacitive sensor variation.

Figure 12.8 shows the signal conversion and conditioning required to produce a control signal from the sensor.

The set point will be a d.c. voltage level. Amplifier A_1 adds the phase sensitive rectifier output to the set point so that both inputs of difference amplifier A_2 are referred to the same level.

The hydraulic servo valve responds to the output of A_2 and allows hydraulic power to the actuator. This then raises or lowers the roller, depending on the polarity of the phase sensitive rectifier output.

Is this the best solution?

Using a capacitive displacement sensor to detect the small variations in the thickness of sheet steel is particularly suited to this application. The displacements are small, the device has infinite resolution, fast response, is sensitive and can cope with the harsh environment.

The closed-loop control system described should be effective in adjusting the rollers to produce steel strip of the required thickness. If necessary a servo chart recorder could be added to the system for quality control purposes.

To condition the signals we have used devices we have not met previously, such as the phase sensitive rectifier and the a.c. bridge circuit. However, the signal conditioning requirements are fairly straightforward and should not present any major difficulties.

There do not appear to be any major problems or uncertainties with the solution at this stage. This outline solution provides a sound basis on which to develop a practical design.

Case study: container filling in a chemical process

Continuous processes, such as those found in the petrochemicals industry, often rely on chemical constituents being mixed in carefully controlled amounts, in a particular order. The amounts have to be precise and accurate, not only to prevent product variation, but also to ensure the end product is of the desired quality. This case study looks at the control of a powder hopper in a continuous process.

What is the problem?

To optimise the quantity of products processed in a given period in a chemical plant, an automatic measuring system is needed. This measuring system is to be in the form of a hopper. This hopper will add controlled amounts of a chemical powder to a mixing process, in a designated order.

The hopper will be filled with powder via an inlet valve, until the powder reaches a set level. When the hopper is filled to this level it will contain the correct amount of powder for the mixing process. When the mixing process needs this powder, the hopper should discharge it via an outlet valve.

Both inlet and outlet valves are electrically operated. To ensure inherently safe operation these valves are of the type that need electrical power to open. Thus if power fails the valves will be closed and thus overfilling or wrong mixing is avoided.

A system is needed to ensure the measuring hopper is filled to exact amounts. The system must then release these amounts into a mixing container at the correct time, as demanded by the mixing process.

Measurement of the amount must be accurate, to about 2%, in order that the synthesised product meets its specification.

This operation is part of a wider process which is monitored and controlled by a computer.

How can we solve it?

We can use sensors in the system which send a control signal to the inlet and outlet valves so that they:

- Close the hopper outlet during filling.
- Close the hopper inlet when the hopper contents have reached their correct amount.
- Open the hopper outlet when the contents need to be discharged.

We are mainly concerned with measuring the hopper contents and then closing the inlet valve when the contents has reached its required value. To achieve this we must measure the level of the contents in the hopper. This must be monitored until it reaches the set level, and then the inlet valve should be closed. Figure 12.9 shows how we can achieve this.

As the hopper fills, the weight of the powder will cause it to move slightly downwards. The hopper can be designed such that this displacement is proportional to the amount of powder in it. Sensors can be mounted at the base of the hopper to detect this displacement.

The control signal needs to be a d.c. voltage with enough current to power the valve operating mechanism.

What sensors could we use?

The hopper contents can be measured in a number of ways. Three of these are

- Proving ring with Gray coded disc.
- Column load cell with strain gauge bridge.
- Beam load cell with strain gauge bridge.

The proving ring consists of a steel tube, modified at the top and bottom so that vertical loads can be applied across its diameter. The ring deflects under load, and this deflection is measured.

In test laboratories, the deflection of a proving ring is measured with a dial test indicator. However, commercial proving rings may detect the distortion using a linear differential transformer, a rotary differential transformer, strain gauges or optical shaft encoders.

We shall employ the shaft encoder. The upper mounting of the proving ring incorporates a rack gear and the lower mounting a bracket to support the optical shaft encoder, as shown in Figure 12.10.

The optical encoder shaft has a pinion gear in mesh with the rack gear. Thus the small linear displacement of the ring can be magnified to a large rotary motion by the shaft encoder gearing. Infra-red light sources and sensors with a Gray-coded disc between them will give a digital output. Eight tracks will be used to give a resolution of 1/256 of a revolution, or 1.4 °.

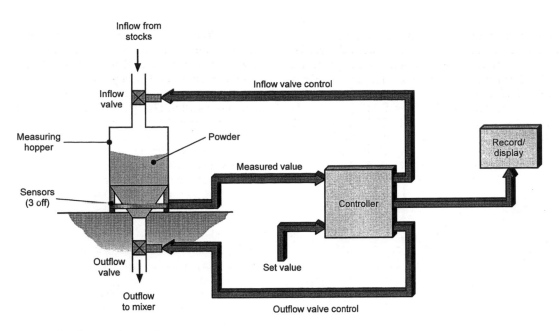

Figure 12.9 Controlling hopper contents

The column load cell consists of a tube or a solid bar loaded axially. The strain in the cell is measured by bonded resistance strain gauges. These are cemented in four equally spaced positions around the circumference of the bar. If the bar is under load it deforms slightly, and this deformation is proportional to the force applied to it. Hence the strain gauges detect the tensile or compressive loads applied to the cell. Using four strain gauges compensates for changes in temperature and, being mounted opposite to each other, the output is insensitive to loads being applied off centre or at an angle.

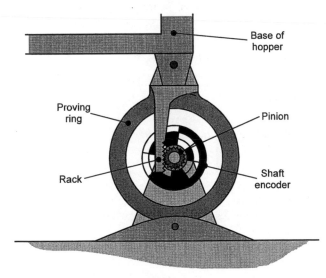

Figure 12.10 Proving ring with shaft encoder

The beam load cell is a hollow piece of tube with a rectangular cross-section (a box section). It has mountings at the top and bottom, so that loads can be applied across two opposite sides. Strain gauges are fixed to one of the sides of the box. The four strain gauges again offer temperature compensation, but the output is somewhat sensitive to off-centre loading.

Both the column load cell with strain gauge bridge and the beam load cell with strain gauge bridge give an analogue output.

They have no moving parts in contact to cause wear and thus require little maintenance.

The optical shaft encoder provides a digital output, which is of a form suitable as an input to a computer recording display and control facility. It has parts that might wear and so would need to be sealed to prevent dirt and dust clinging to the moving parts. There is friction between the rack and pinion and the bearings of the shaft encoder, so maintenance would be required.

All three examples require low voltage external power supplies. We shall choose the shaft encoder because its digital output is suited to the computer control of the wider process. The other two sensors would require analogue to digital conversion of their outputs. Although less accurate than the other two, it will be within the 2% accuracy required by this application.

What type of display is most suitable?

Control of the hopper contents forms part of a wider process in which a number of constituents are combined in specified proportions. There are other operations involved, such as heating as well as mixing.

The overall process is monitored and controlled by a computer based system. This allows various displays to be provided. These show the hoppers filling and emptying, mixing and heating taking place, and the rate at which the final product is being manufactured. The process operator can switch between different screen displays to observe and monitor a range of different variables and control loops.

Use of the shaft encoder will allow digital values to be input directly to the computer and readily:

- Recorded for later analysis.
- Processed for display.
- Used to generate control signals.

The set point value can be entered via the computer keyboard and shown as a bar graph display. The display will also show the measured value and the inlet and outlet valve conditions. This screen display would be as shown in Figure 12.11.

Open condition highlighted

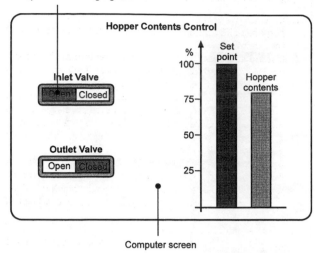

Figure 12.11 Real-time display of contents and valve states

When the increasing hopper contents bar equals the set point value bar, the valves will activate and the display indicators will change over.

The relative values of the set point and measured value over a series of fillings will be recorded in the computer memory. This will allow recent trends to be displayed as an aid to ensuring quality control of the process.

How do we condition the signals and link the devices?

The hopper is supported on three proving ring sensors, the outputs of which must be added together to determine the actual contents. Each proving ring has a shaft encoder with a Gray-coded disc. The Gray codes must be converted to binary values before being

added. This can be achieved either by electronic hardware decoders or by computer software.

If we use computer software, there is very little electronic signal conditioning or processing required. Light sources and detectors are provided for each shaft encoder and the data can be read into the computer via a 3-way multiplexer. Figure 12.12 shows this arrangement.

Each sensor output is connected to an 8-bit data register. These hold the sensor output value in Gray code form. The computer software programme samples the sensor values in sequence.

The arrangement shown in Figure 12.12 forms a 3-way multiplexer. This allows the three sensors to share one computer port. If three computer ports are available the multiplexer would not be required and the sensors could be connected directly.

Each Gray code value read by the computer is matched to a table of binary values, located in the computer's memory. The correct value is chosen and added to the corresponding values from the other two sensors. This produces the bar graph display height and checks if the set point is reached. If a match is found the valve control signal will be sent to close the inflow. The outflow valve can then be opened at some later time, as required by the chemical process.

Is this the best solution?

The sensor output is in digital form, providing level sensing of the required accuracy for the chemical mixing process. Signal conditioning is fairly straightforward because computer control is already in place. The solution provides an outline to build on.

There are many other sensors we could have considered to measure the level of the chemical powder in the tank. Most of these would require analogue to digital signal conversion. However, unless we explore these options we cannot be sure we have arrived at the best solution. There may be more efficient, accurate, reliable or less expensive possibilities we have overlooked. Therefore, even though the solution looks sound, further investigation is necessary.

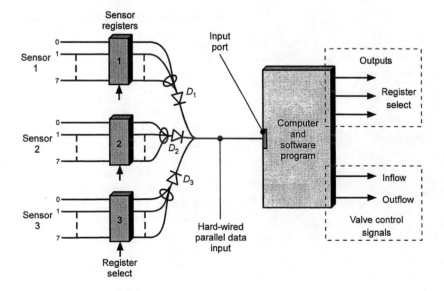

Figure 12.12 Reading in sensor data for processing, display and control

Summary

In this chapter we looked at three case studies which use sensors in control situations. The first case study used open-loop control, the other two closed-loop. Purely open-loop systems do not always need to incorporate sensors. They rely on the set value being appropriate to the system under control and so do not need to sense or measure the output. From the first case study we can see that the sensors incorporated in an open-loop system are for measurement purposes, not control.

In a similar manner to the measurement applications in Chapter 11 we have used a methodical approach to find the most suitable choice of sensor for the control application, and outlined possible solutions to the control problem. After discussing these solutions we have commented on their suitability and decided whether further investigation is needed before producing a detailed design. Control applications are often more complex than pure measurement applications, often forming part of a larger process. For the best choice of sensor to be made as much information about the control problem, and all possible solutions should be taken into account.

Questions for further discussion

1. In what ways does the choice of sensor for control purposes differ from the choice of sensor for measurement purposes?
2. List the advantages and disadvantages that might occur if closed-loop control of the conveyor belt speed was implemented in the first case study.
3. Even though reed-switches are relatively cheap, why might the overall cost of using a system incorporating them to measure the conveyor belt speed in the first case study work out more expensive than the a.c. tachometric generator solution chosen?
4. In the steel mill case study, the thickness of the steel is measured after it has been rolled. If it is out of tolerance rework will be necessary. Discuss ways in which you think other sensors could be employed to reduce this level of rework.
5. What sensors, other than those discussed, could be used to detect the level of the powder in the hopper in the chemical process case study?
6. List the possible advantages and disadvantages of using open-loop control to fill the hopper in the chemical process case study.

13 Practical Experiments

This chapter contains a series of experiments supporting the theory given previously. Ideally they should be performed on the TecQuipment Sensors and Instrumentation System hardware. However, it may be possible to perform the experiments using similar components to those described here. Because the specifications of these components may vary, results are not included. Figure 13.1 shows the TecQuipment Sensors and Instrumentation System hardware.

The experiments assume a knowledge of basic electrical engineering principles. To gain full benefit from these experiments, read and understand at least Chapter 2, Chapter 3, Chapter 9 and Chapter 10.

Each experiment has a number of questions at the end. After completing the experiment, try and answer these, even if you need to take further measurements or revise the topic. You may also benefit from using the results obtained, observations made and the answers given to the questions to write a report on each experiment. In the report, include any theory you feel supports the comments and conclusions you give. If possible, suggest any changes to the experiment that would improve the quality of the results and widen the scope of the experiment.

For all experiments, switch on the equipment for a minimum of 5 minutes prior to any settings being made or any results taken. This will ensure that any drift in the circuits due to thermal effects will be kept to a minimum.

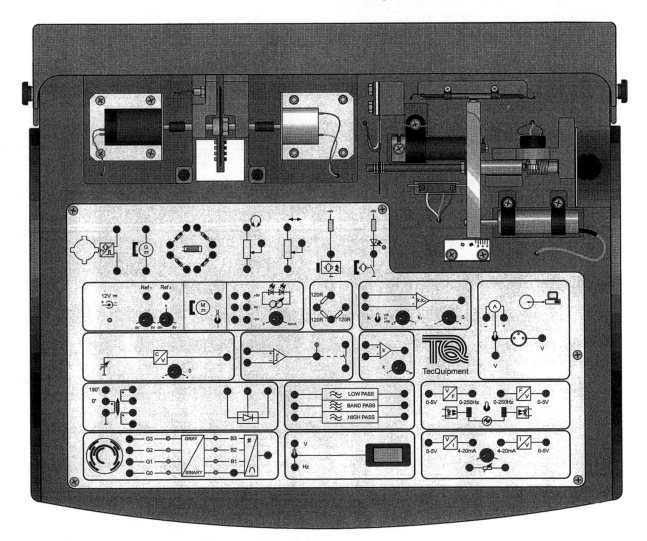

Figure 13.1 TecQuipment sensors and instrumentation hardware

Sensor specifications and familiarisation

This section describes the sensors, signal conditioning equipment and power supplies needed to perform the experiments. They are as on the TecQuipment SIS hardware. If you intend to perform the experiments on other equipment, it should be of a similar specification.

To gain the full benefit of the experiments, it is worthwhile spending some time becoming familiar with each element. Consider how the components are constructed and positioned within a measurement system, and how they respond to changes in the measurand. In later experiments we will be returning to these circuits and devices and investigating them in more detail as well as putting them to practical use.

The linear assembly

Figure 13.2 shows the linear assembly. It is used for experiments which study sensing linear motion. It consists of rotary scale which produces a linear motion which can be detected by various sensors connected to it. This linear motion can be left or right, by manually turning a scaled control, shown in Figure 13.3.

The rotary scale of the linear assembly is scaled 0 to 0.9. One complete revolution of the scale moves the linear assembly ±1 mm. Clockwise moves it to the right and anticlockwise to the left. The rotary scale may therefore be used to indicate linear movement with a resolution of ±0.05 mm. The maximum range of movement possible with the linear assembly is 9 mm or ±4.5 mm.

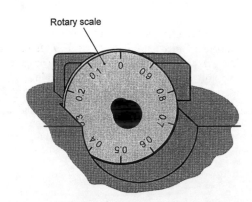

Figure 13.3 Rotary scale

The range of movement of the linear assembly is limited at each end of its travel. Do not apply any undue force to move the assembly beyond these limits. Permanent damage to the equipment may result if excess force is used.

The measurement interval used in all the later experiments on linear motion is 1 mm. This is one complete revolution of the rotary scale. The following sensors are mounted on the linear assembly.

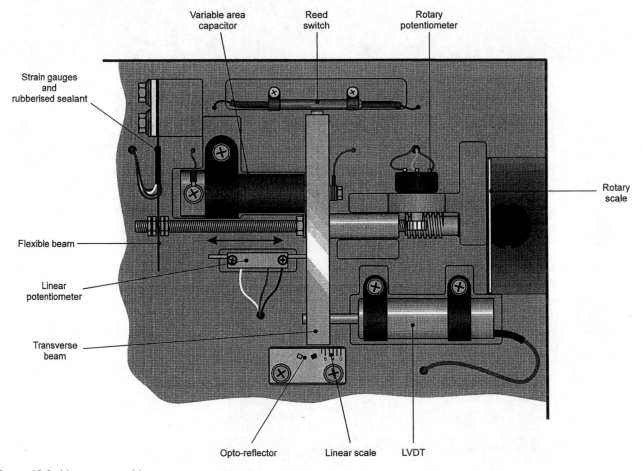

Figure 13.2 Linear assembly

Strain gauges and the flexible beam assembly

This is a sub-assembly of the linear assembly. It comprises a flexible beam which is deflected when the linear assembly is moved. Four foil type bonded resistance strain gauges are attached to the flexible beam. Two are tensioned when the beam is strained in one direction, because they are on the outside of the curvature of the beam. The other two are in compression when the beam is moved in the same direction, because they are on the inside of the curvature of the beam. When the beam is strained in the opposite direction this action reverses.

The strain gauges have a resistance of 120 Ω and a gauge factor: 2.12 (nominal). For protection the they are coated with a transparent rubberised sealant. Figure 13.4 shows the section of the front panel which accesses the strain gauges.

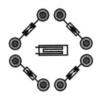

Figure 13.4 Strain gauge section of the front panel

Linear potentiometer

The linear potentiometer has a total resistance of 10 kΩ. Its maximum displacement (stroke) is 10 mm. It has a ±0.4% linearity and a ±10% tolerance.

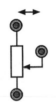

Figure 13.5 Linear potentiometer section of the front panel

Figure 13.5 shows the section of the front panel which accesses the linear potentiometer.

Rotary potentiometer

The rotary potentiometer is rated at 10 kΩ. The resistor material is conductive plastic. It has a linearity of ±2% and a tolerance of ±20%. It rotates 340° (or ±170°) through a worm and wheel gearing arrangement to the linear assembly. The gear ratio produces 0.0833 (one twelfth) of a revolution of the rotary potentiometer shaft for each complete rotation of the calibrated scale. For the full 9 mm movement of the linear assembly the shaft of the rotary potentiometer will rotate through 270°. Figure 13.6 shows the section of the front panel which accesses the rotary potentiometer.

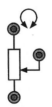

Figure 13.6 Rotary potentiometer section of the front panel

Linear variable differential transformer (LVDT)

The LVDT has six access wires, two for the primary winding and two for each of the secondary windings. It requires a 3 kHz, 5 V supply. Its linearity is ±0.5% and sensitivity ±75 mV/mm.

The plunger of the LVDT is rigidly attached to the linear assembly. The body of the LVDT is fixed to the chassis of the hardware module. When the linear assembly is moved there is relative motion between the body of the LVDT and the link connecting to its soft iron core.

The supply to the primary winding and output signals from the two secondary windings are available at the phase sensitive detector block on the front panel (see later).

Variable area capacitor

A variable area capacitor comprising an outer cylinder attached to the chassis of the hardware module, and a second sliding cylinder attached to the linear assembly. They are mounted and aligned such that the inner cylinder is partially inserted into the outer cylinder. The inner cylinder is covered with a plastic material to act as an insulator and also as a dielectric.

When the linear assembly is moved to the left, the inner cylinder enters the static outer cylinder to increase the area of overlap and decrease it when moved to the right.

The variable area capacitor is accessed via the capacitance to voltage converter.

Reed switch

The reed switch is the normally open type. It consists of rhodium contacts mounted in a hermetically sealed glass tube. The contacts close or open when a magnet, mounted in the end of the adjacent transverse beam, is moved to be in close proximity and then moved away. An LED illuminates when the contacts are closed. The circuit must be complete for current to flow through the LED to cause it to illuminate.

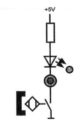

Figure 13.7 Reed switch section of the front panel

Figure 13.7 shows the section of the front panel which accesses the reed switch. It is rated 80 VA, 1.3 A d.c. or a.c. maximum, switching voltage 250 V a.c. rms., pull-in range 40 to 45 AT, breakdown voltage 800 V d.c. The resistor limits the maximum amount of current which can flow through the switch.

Reflective optical beam sensor

This sensor comprises a light emitting diode (LED) and a phototransistor. This is mounted adjacent to the reflective surface of the end of the transverse beam of the linear assembly. Changes in position of the linear assembly affect the amount of light reaching the phototransistor.

An adjacent scale, marked in 1 mm steps, gives an approximate indication of the linear assembly position and relative movement. For accurate measurements always use the rotary scale though be sure not to lose count of the number of turns.

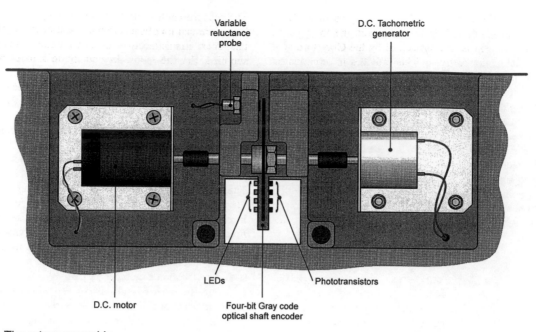

Figure 13.8   The rotary assembly

The opto-reflector shield allows investigations into the effect of background levels of light and how this affects the sensitivity of the system. Figure 13.9 shows the section of the front panel which accesses the rotary potentiometer.

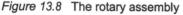

Figure 13.9   reflective optical section of the front panel

The rotary assembly

The rotary assembly comprises a horizontal shaft with a number of sensors attached, as shown in Figure 13.8. These are for measurement of rotational speed and position. The d.c. motor drives the shaft at various selected speeds, from 0 to 209.5 rad.s^{-1} (0 to 2000 revolutions per minute). The output voltage or frequency of the rotational sensors may be measured using the panel mounted digital voltmeter (DVM) to provide the information against which to calibrate the individual speed sensors.

Speed may be set manually or automatically using closed-loop control with one of the speed sensors providing the feedback signal.

With Ref$_2$ connected to the motor input and the selector switch set to '1' the speed of the motor may be varied in terms of both speed and direction of rotation. Setting the selector to '0' disables the motor supply.

Access to the d.c. motor may be made at the appropriate socket located in the power supply section (see later).

D.C. tachometric generator

A permanent magnet d.c. tachometric generator producing a nominal output of 2.5 ±0.25 V per 104.7 rad.s^{-1} (1000 revolutions

per minute). With the motor driven at different speeds the output from the tachometric generator may be displayed on the digital volt meter. The polarity indicates the direction of rotation. Figure 13.10 shows the section of the front panel which accesses the d.c. tachometric generator.

Figure 13.10   The d.c. tachometric generator

Four-bit encoder

The four-bit encoder consists of metal disc with a four-bit Gray scale formed by a series of slots. Four pairs of LEDs and phototransistors are mounted on either side of the disc, as illustrated in Figure 13.11.

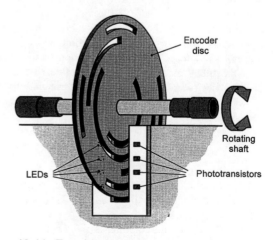

Figure 13.11   Four-bit encoder

The sequence of 'highs' and 'lows' at the outputs of the four phototransistors are indicated by four LEDs on the front panel. This digital information is then processed by the Gray to digital converter into an analogue signal. This provides information on the angular position of the disc, its speed of rotation and direction of rotation.

An indexing mark on the edge of the encoder disc indicates position zero when all bits are low. Access to the four-bit encoder is via the decoder section of the front panel (see later).

Optical tachometer

The optical tachometer uses the least significant bit (LSB), of the encoder to produce a pulsed output. The frequency of this output is directly proportional to the rotational speed of the shaft.

Access to the optical tachometer may be made in the decoder section of the front panel (see later).

Variable reluctance probe

The variable reluctance proximity sensor is an encapsulated sensor comprising a steel outer body supporting a permanent magnet with a coil wound around it. Each complete revolution of the rotary disc produces four pulses at the output terminals.

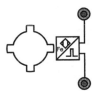

Figure 13.12 The variable reluctance probe

Figure 13.12 shows the section of the front panel which accesses the variable reluctance probe.

The signal conditioning circuits

The signal conditioning circuits on the hardware module are a selection of electronic circuits designed to condition raw signals from the sensors into a suitable form. This is usually a d.c. voltage level for data acquisition or control purposes.

Differential amplifier

This is a dual input differential amplifier with variable gain and set zero controls. Figure 13.13 shows the section of the front panel which accesses the differential amplifier.

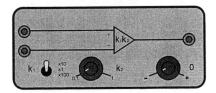

Figure 13.13 The differential amplifier

The range switch allows gain k_1: ×1, ×10, ×100. The gain k_2 is continuously variable between 0.1 and 1.0. The amplifier gain is the product of these two settings, namely $k_1 \times k_2$, giving an overall range of gains of 0.1 to 100.

The output voltage of this amplifier is given by:

$$V_{out} = (V_{+ve} - V_{-ve}) \times k_1 \times k_2 + V_{off}$$

The set zero control offsets the output signal in the range ±5 V. To ensure that no offsets are present, it must be calibrated. To do this, short circuit the two inputs with one of the patching leads supplied. Set the gains, k_1 and k_2, to minimum. Observe the output signal using the DVM set to read volts (V). If there is an offset, adjust the set zero until it is removed. Increase the gains to maximum and zero the output of the differential amplifier as indicated by the DVM.

If the 'set zero' control is moved it will be necessary to repeat the calibration procedure.

Capacitance to voltage converter

This is an a.c. capacitance to voltage converter with a built in 80 kHz oscillator. An external capacitance connected across the input forms one arm of an a.c. bridge. A set zero control is included to reduce any offset in the output signal to zero. Figure 13.14 shows the section of the front panel which accesses the capacitance to voltage converter.

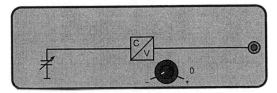

Figure 13.14 The capacitance to voltage converter

The supply to the capacitor cannot be measured. The oscillator is only used with the capacitor and the permanently wired connections are made with screened leads to minimise noise. The output voltage is directly proportional to the capacitance.

Comparator

The comparator provides a comparison between two signals such that the output is either 'high' or 'low'. This depends on which is more positive. An LED at its output indicates when the output is 'high'. Figure 13.15 shows the section of the front panel which accesses the comparator.

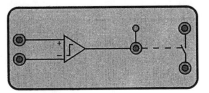

Figure 13.15 The comparator

If the voltage on the positive input is greater than the voltage on the negative input, the output voltage will be high. At all other times the output voltage will be low. The output allows the comparator to control the supply to devices such as the d.c. motor or compatible external device. The supply is either on or off depending upon the relative magnitudes of the input signals.

Phase sensitive detector

The phase sensitive detector is dedicated for use with the LVDT. It produces a d.c. output signal proportional to the phase of the input compared to a reference signal. Figure 13.16 shows the section of the front panel which accesses the phase-sensitive detector.

Figure 13.16 The phase-sensitive detector

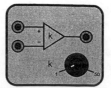

Figure 13.19 The summing amplifier

It includes a 5 V (peak) 3 kHz oscillator to energise the LVDT primary winding as well as providing 0° and a 180° references.

Figure 13.19 shows the section of the front panel which accesses the summing amplifier.

Four-bit decoder

The four-bit decoder is used with the four-bit Gray scale encoder. It gives rotary position measurements with a resolution of 22.5° (360°/16). Figure 13.17 shows the section of the front panel which accesses the four-bit decoder.

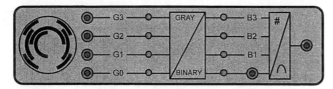

Figure 13.17 The four-bit decoder

The decoder inputs the four individual bits of data directly from the Gray scale encoder. The status of each bit is indicated by an LED. The decoder converts these into a binary code using a Gray to binary converter, again the status of each bit being indicated by an LED. The binary code is then converted by a four-bit digital to analogue converter (DAC) into an analogue signal in the range of 0 to 5 V. This corresponds to a complete revolution of the shaft. With the four-bit resolution of the encoder this signal will comprise a series of 16 steps corresponding to angular positional changes of 3.6 radians (22.5°).

The disc may be manually rotated to observe the effect on the output states, as indicated by the LEDs. The variable current source located in the power supply section may be used to vary the level of illumination of the LEDs to investigate performance under varying levels of ambient light.

Filters

The filters remove unwanted frequencies from a signal. Filter options are: low pass, band pass and high pass. Figure 13.18 shows the section of the front panel which accesses the filters.

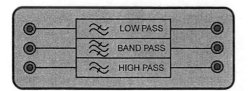

Figure 13.18 The filters

Summing amplifier

This is a differential amplifier like the one described earlier. The summing amplifier provides an error signal for control applications. amplifier with one non-inverting (+) and one inverting (–) input. The output is the difference between the two inputs. A rotary gain control varies the gain of the amplifier in the range of 1 to 50.

Voltage to frequency (V to F) and frequency to voltage (F to V) converters

The V to F converter converts an input signal in the range 0 to 5 V into a corresponding frequency of 0 to 250 Hz. The F to V converter converts a 0 to 250 Hz signal at its input into a 0 to 5 V signal at its output. The connection between the V to F and F to V circuits is made using a patching lead. The V to F and F to V circuits may be used together or separately.

An optical link between the V to F and F to V circuits, selected by an adjacent toggle switch, allows the frequency modulated signals to be transmitted through a fibre optic cable connecting the two converters as an alternative to a conductor. An LED and phototransistor at the sending and receiving end provide the electrical to light and light to electrical conversions.

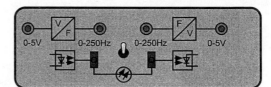

Figure 13.20 The voltage to frequency and frequency to voltage converters

Figure 13.20 shows the section of the front panel which accesses the V to F and F to V converters.

The output frequency from the V to F converter is given by:

$$F_{out} = V_{in} \times 50$$

The output voltage from the frequency to voltage converter is given by

$$V_{out} = F_{in} \times 50$$

Voltage to current (V to I) and current to voltage (I to V) converters

The V to I and I to V converter illustrates data transmission using current sources to overcome losses in a cable when using analogue voltage signals. A voltage input to the V to I converter in the range 0 to 5 V produces a current output of 4–20 mA. A current input to the I to V converter in the range 4 to 20 mA produces a voltage output signal in the range of 0 to 5 V. Figure 13.21 shows the section of the front panel which accesses the V to I and I to V converters.

A 0 to 500 Ω variable resistance simulates a length of cable between the source and the receiver. The converters can be used to send and receive data using external sensors, with a maximum resistance of approximately 300 Ω.

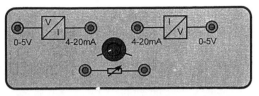

Figure 13.21   The voltage to current and current to voltage converters

Bridge completion resistors

The bridge completion resistors are set of three 120 Ω precision resistors used with one or more of the strain gauges. Figure 13.22 shows the section of the front panel which accesses the bridge completion resistors.

Figure 13.22   The bridge completion resistors

Panel meter

The panel meter is a three-and-a-half-digit digital voltage and frequency meter. Its ranges are 0 to ± 10 V in 10 mV steps; or 0 to 500 Hz, accuracy ±1%. Figure 13.23 shows the section of the front panel which accesses the panel meter.

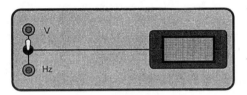

Figure 13.23   The panel meter

PC interface

The PC interface connects the hardware to a PC if data acquisition software is used. It consists of a twelve bit analogue to digital and digital to analogue converter with a sampling frequency up to 10 kHz. It provides one analogue to digital (A to D) input and one digital to analogue (D to A) output. Figure 13.24 shows the section of the front panel which accesses the PC interface.

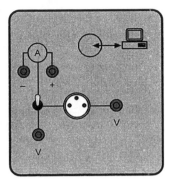

Figure 13.24   The PC interface

The input is switchable for a voltage input in the range ±5 V, with a resolution of ±2.5 mV, or a current input in the range ±500 mA, with a resolution of ±0.25 mA.

Power supply section

This section provides the supplies needed by the sensors, signal processing circuits and the d.c. motor. The supplies available are: variable 0 to 5 V (Ref₁); variable ±5 V d.c. (Ref₂); and fixed d.c. supplies of +5 V, −5 V and 0 V (two of each). It also includes a variable 0 to 30 mA current source for the LED sensors in the encoder and reflective optical beam sensors. Figure 13.25 shows the section of the front panel which accesses the power supplies.

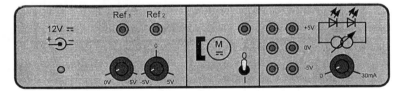

Figure 13.25   The power supplies

Experiment 1: the strain gauge

The objective of this experiment is to assess the performance of strain gauges measuring linear displacement. On completion of this experiment you will:

- Appreciate the positioning of strain gauges measuring displacement.
- Understand the use of strain gauges in a potential divider, ¼, ½ and full bridge configurations and the relative sensitivities in each case.
- Assess the sources of errors in using strain gauges to measure displacement.

Part (a): strain gauge potential divider

The circuit shown in Figure 13.26 is a simple potential divider with a strain gauge (tension), R_{sg}. It is connected in series with a fixed matching resistor, R, and a power supply connected across both, V_{in}.

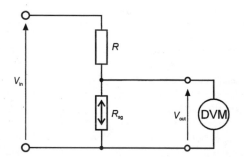

Figure 13.26 Strain gauge potential divider

The equation relating the output voltage to the other circuit parameters is:

$$V_o = V_{in} \frac{R_{sg}}{R_{sg} + R}$$

In other words, the output is governed by the ratio of the strain gauge resistance, R_{sg}, to the total resistance, $R_{sg} + R$.

Using patching leads connect the equipment as shown in Figure 13.27.

Move the linear assembly to the right by turning the rotary scale clockwise until it reaches the end stop. Carefully adjust the dial until the zero aligns with the edge of the moulding. Record the initial value in volts of the meter reading.

Compare this reading with the value calculated from the equation given. Was the experimental value expected?

Turn the rotary scale anticlockwise to move the linear assembly over the whole range and observe the change in the indicated voltage. Return the linear assembly to the start position once you have completed this exercise.

You should see that the indicated meter reading changes very little, if at all. However, we know that the strain gauge resistance has to have changed with the displacement.

The change in resistance is very small and the corresponding voltage change is less than the resolution of the meter (0.01 V). We need to amplify the output signal to increase the sensitivity of the measurement system.

Modify the circuit to include the differential amplifier between the output and the meter, as shown in Figure 13.28 and represented schematically in Figure 13.29.

Set the gains of the amplifier, k_1 and k_2, to maximum. Adjust the setting of Ref_1 to make the indicated meter value as small as possible (less than 5 V to be within the range of the 'set zero' control) and then zero the reading with the 'set zero' control.

Note that with the gain controls of the differential amplifier set to maximum (gain = 100) the adjustment is very coarse. A 1 mV difference between the two input voltages will be enough to cause the meter reading to change from +10 V to −10 V.

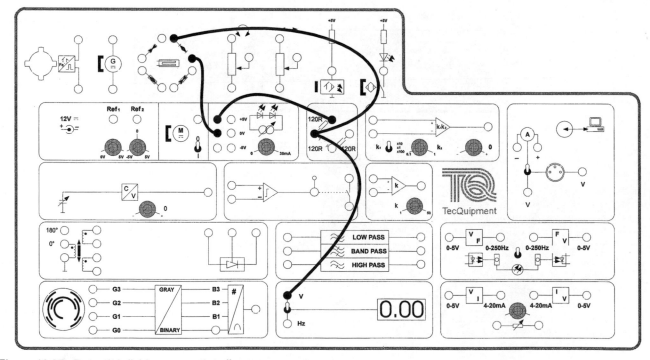

Figure 13.27 Potential divider connection diagram.

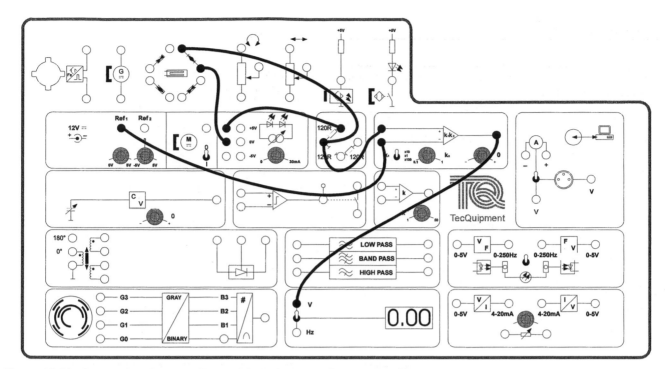

Figure 13.28 Connection diagram of potential divider with differential amplifier

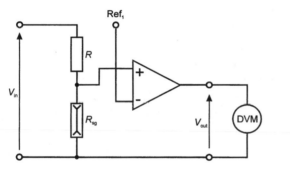

Figure 13.29 Strain gauge potential divider with differential amplifier

Displacement (mm)	Output (V) left	Output (V) right
0		
1		
2		
3		
4		
5		
6		
7		
8		
9		

Figure 13.30 Results table for strain gauge experiments

The output signal is now the amplified difference between the two inputs signals, the potentiometer value and that of Ref_1. The change in the output will only be caused by the changes in potential divider ratio due to the changes in R_{sg}.

In steps of 1 mm (one complete revolution of the rotary scale) move the linear assembly to the left over its full range of travel and record corresponding meter readings to complete the table given in Figure 13.30. Be careful to adjust the control in one direction only for each set of readings.

Plot a graph of your results. Determine the sensitivity of the measurement system from the slope of the graph and also the intercept with the vertical axis. Hence determine the equation relating the meter reading with displacement.

Comment on the linearity, hysteresis, scatter and repeatability of the measurements obtained.

Over the same range of displacements with one of the compression gauges replacing the tension gauge, repeat the previous procedure and tabulate your results.

Plot your results and compare them with those obtained with the tension gauge.

Part (b): the quarter strain gauge bridge (or quarter bridge)

This experiment determines the performance of the quarter bridge and compares it with the results of the potential divider obtained previously.

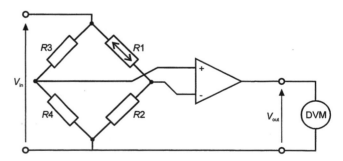

Figure 13.31 Quarter strain gauge bridge with differential amplifier

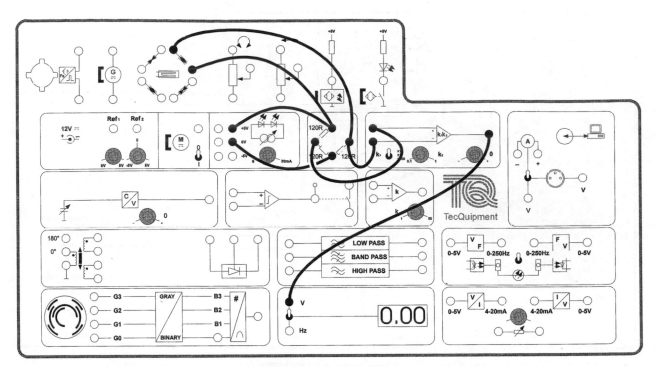

Figure 13.32 Connection diagram of quarter bridge with differential amplifier

The quarter strain bridge is a conventional bridge circuit which is an example of a Wheatstone bridge. One arm is formed by a strain gauge in series with a fixed resistor ($R_1 + R_2$) and the other by two fixed resistors ($R_3 + R_4$). The quarter strain gauge bridge with a differential amplifier is shown schematically in Figure 13.31.

The strain gauge arm produces a change in output with change in strain while the resistor arm produces a fixed voltage to replace that produced previously with the additional, variable, supply (Ref₁). Effectively the performance should be the same as for the potential divider circuit in Part (a) but at a lower cost.

Connect the quarter bridge strain gauge circuit as shown in Figure 13.32. Repeat the procedure described in Part (a) and tabulate your results.

Plot a graph of your results. From this determine the sensitivity of the measurement system from the slope of the graph and also the intercept with the vertical axis. Hence determine the equation relating meter reading with displacement.

Compare your results with those obtained in Part (a) for the potential divider circuit.

Part (c): the half strain gauge bridge (or half bridge)

This experiment determines the performance of the half bridge and compares it with the results of the potential divider and quarter bridge obtained in Parts (a) and (b).

The half bridge is a further enhancement to the basic Wheatstone bridge. One arm is formed by two strain gauges (R_1 and R_2), one positioned to experience increasing tension and the other increasing compression when the linear assembly moves in one direction.

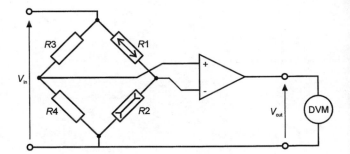

Figure 13.33 Half strain gauge bridge with differential amplifier

Figure 13.33 is a schematic drawing of the half strain gauge bridge, with a differential amplifier. The variation in output is affected by the change in resistance of both strain gauges. The fixed resistor arm (R_3 and R_4) has the same function as before, that is, to produce a fixed reference voltage with which to compare the variable output of the strain gauge arm.

Connect the circuit as shown in Figure 13.34. Repeat the previous procedure and tabulate your results. From this, plot a graph.

Determine the sensitivity of the measurement system from the slope of the graph and also the intercept with the vertical axis. Hence determine the equation relating the meter reading with displacement.

Compare your results for the half bridge with those obtained for the potential divider and quarter bridges in Parts (a) and (b).

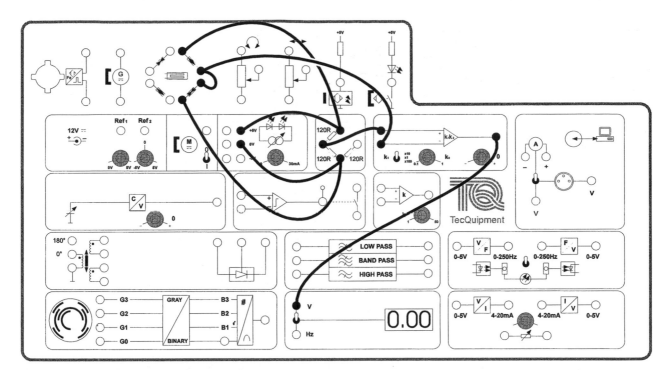

Figure 13.34 Connection diagram of half bridge with differential amplifier

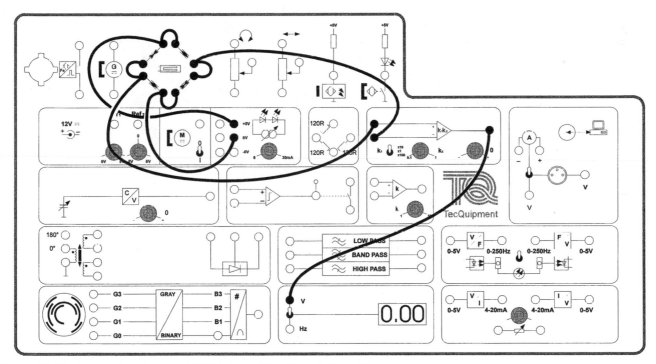

Figure 13.35 Connection diagram of full bridge with differential amplifier

Part (d): the full strain gauge bridge (full bridge)

This experiment determines the performance of the full bridge and compares it with the results of the potential divider, quarter and half bridge obtained previously.

The final development to the Wheatstone bridge circuit is to replace all fixed resistors with strain gauges: two tension and two compression. When the linear assembly is moved, the output voltage from one arm increases (becomes more positive) while the other reduces (becomes less positive). The difference between these two signals, as supplied to the amplifier, increases the overall measured signal amplitude at the output.

Connect the circuit as shown in Figure 13.35. Repeat the previous procedure and tabulate your results. From this, plot a graph. Determine the sensitivity of the measurement system from the slope of the graph and also the intercept with the vertical axis. Hence determine the equation relating the meter reading with displacement

Compare your results with the potential divider, quarter bridge and half bridge investigated in Parts (a), (b) and (c). Figure 13.36 shows the full strain gauge bridge schematically.

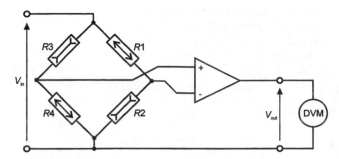

Figure 13.36 Full bridge strain gauge with differential amplifier

Questions

Answer the following questions, making additional measurements if necessary. It may be helpful to refer to the theory given in Chapters 3, 9 and 10.

1. What is the smallest movement of the linear assembly detected by a change in the meter reading?

2. What is the effect of reversing the polarity of the supply to the potential divider?
3. What is the effect of reducing the voltage supplied to the potential divider?
4. Is there a dead-zone in the measurement system and if so what is its cause?
5. What are the main parameters affecting the quality of the measurement system?
6. What was the effect of replacing the tension gauge with the compression gauge in Part (a)?
7. For the potential divider circuit in Part (a), what is the smallest movement of the linear assembly detected by a change in the meter reading?
8. What is the effect of swapping the strain gauge and the resistor of the quarter bridge?
9. What is the effect of reversing the connections to the differential amplifier?
10. What is the effect of starting measurement with the linear assembly zeroed at the far left and moved to the right?
11. Describe how you could modify this experiment, circuit and procedures to measure temperature using bridge techniques.

Experiment 2: the linear and rotary potentiometer

On completion of this experiment you will:

• Understand how linear and rotary potentiometers may be attached to a system to measure displacement.

• Have produced a calibration graph of both types of device and make judgements on linearity, repeatability, accuracy and sensitivity.

• Appreciate the sources of errors in potentiometric circuits.

Part (a): the linear potentiometer

This part of the experiment investigates the linear potentiometer measuring linear displacement.

Make the connections shown in Figure 13.36. This corresponds to the schematic in Figure 13.37.

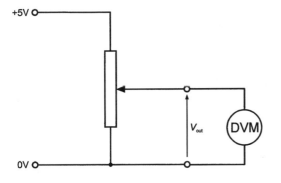

Figure 13.37   Linear potentiometer

Move the linear assembly to the right by rotating the manual control clockwise until it reaches the end stop. Carefully adjust the dial until the zero aligns with the edge of the moulding.

In steps of 1 mm (one complete rotation of the rotary scale) move the linear assembly to the left over its full range of travel. Record corresponding meter readings to complete the table in Figure 13.39. Adjust the control in one direction only throughout the procedure.

Displacement (mm)	Output (V)
0	
1	
2	
3	
4	
5	
6	
7	
8	
9	

Figure 13.39   Results table for potentiometer experiments

Plot a graph of your results. Comment on the shape of the graph and measure its slope and intercept with the vertical axis. Hence give the equation which governs this measurement system.

With the linear assembly in mid position, determine the minimum amount of movement (the resolution) that can be detected by the meter.

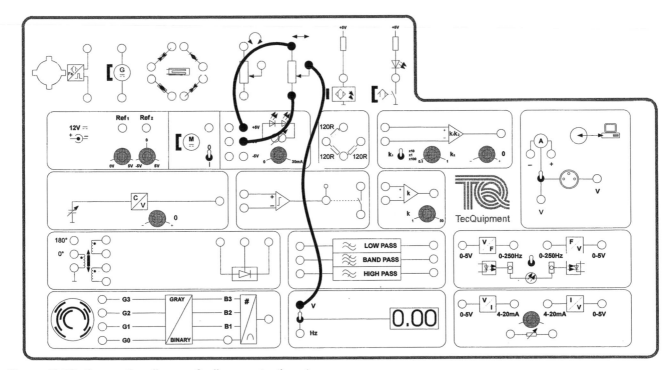

Figure 13.38   Connection diagram for linear potentiometer

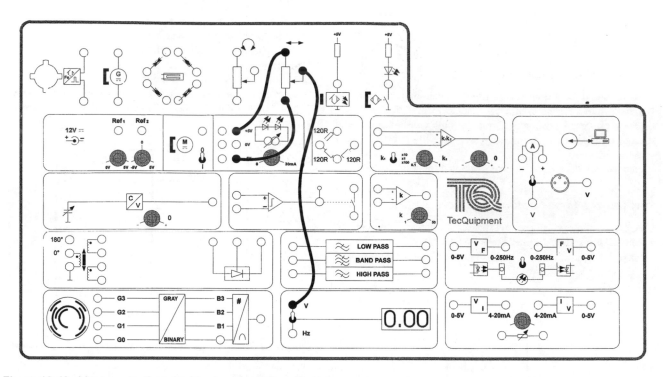

Figure 13.40 Linear potentiometer bipolar connection diagram

Connect the equipment as shown in Figure 13.40. This corresponds to the schematic diagram shown in Figure 13.41. Notice the supply is bipolar, from −5 V to +5 V.

remove any offset in the output signal when the linear assembly is in the starting position.

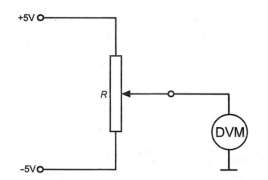

Figure 13.41 Linear potentiometer with bipolar supply

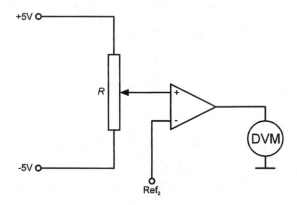

Figure 13.42 Linear potentiometer with differential amplifier

Repeat the previous procedure moving the linear assembly from the right to the left. Tabulate corresponding values of displacement and meter reading. Plot a graph of your results.

Comment on the shape of the graph and measure its slope and intercept with the vertical axis. Hence give the equation which governs this measurement system.

With the linear assembly adjusted to be in mid position determine the resolution of the system.

The circuit shown in Figure 13.42 shows the output from the potentiometer connected to an input (+) of a differential amplifier. An external reference voltage, Ref_2, is connected to the other input (−). The object here is to use the reference voltage to

Connect the equipment as shown in Figure 13.43. This corresponds to the schematic in Figure 13.42.

With the amplifier gain set to unity, adjust Ref_2 so the meter reading is zero at the starting position.

Repeat the previous procedure and tabulate your results. From this. plot a graph.

Comment on the shape of the graph and measure its slope and intercept with the vertical axis. Hence give the equation which governs this measurement system.

With the linear assembly adjusted to be in mid position determine the resolution of the system.

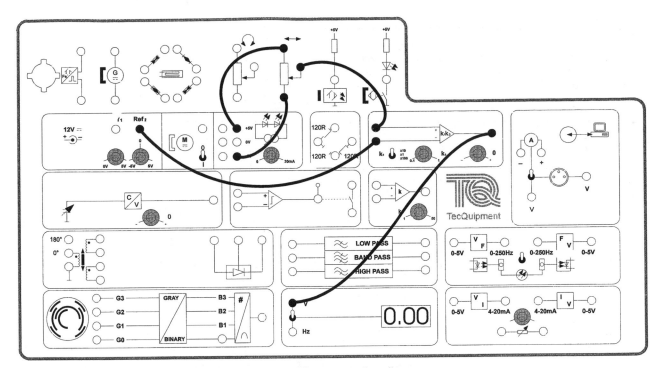

Figure 13.43 Linear potentiometer with differential amplifier connection diagram

Part (b): the rotary potentiometer

The rotary potentiometer illustrates the conversion of linear motion into rotary motion, in this example by the use of a worm and wheel arrangement. Although the shaft is caused to rotate by the movement of the linear assembly the final motion delivered to the potentiometer is actually rotary.

While moving the linear assembly over its full range of movement visually inspect the effect this has on the shaft of the rotary potentiometer in terms of the angle moved through. Also

observe any relative movement between the worm and wheel arrangement which would cause errors in measurement.

Connect the equipment as shown in Figure 13.44. The schematic for this circuit is as Figure 13.42.

Repeat the previous procedure and tabulate your results. Plot a graph and comment on its shape. Measure the slope of the graph and intercept with the vertical axis. Hence give the equation which governs this measurement system.

With the linear assembly adjusted to be in mid position determine the resolution.

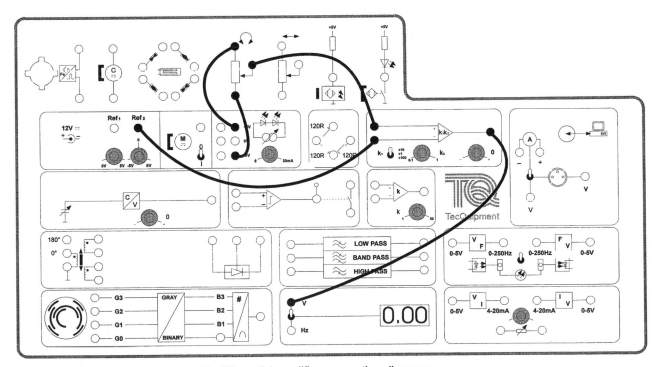

Figure 13.44 Rotary potentiometer with differential amplifier connection diagram

Questions

Answer the following questions, making additional measurements if necessary. It may be helpful to refer to the theory given in Chapters 3, 9 and 10.

1. In the experiments, a variable reference voltage was used to zero the output signal at the start position. How else could zeroing the output voltage at the starting point have been achieved? Hint: Not all solutions need to be electrical.

2. In the SIS, a worm and wheel arrangement is used to convert linear motion into rotary motion. Describe, with the aid of diagrams, two other methods which could achieve this same effect.

3. For each experiment, for both the linear and rotary potentiometers, none gave an output voltage equal to the actual supply voltage. Why should this be the case?

4. Could the rotary potentiometer be used to make measurements of rotary displacements greater than 360°?

5. If the meter used to display the output signal of the circuit shown in Figure 13.38 has an impedance of 5 kΩ, what effect would this have on the output signal when the wiper is at the centre point of the resistance (equidistant from points A and B)? Refer to the specifications for the linear potentiometer given earlier.

Experiment 3: the linear variable differential transformer (LVDT)

On completion of this experiment you will,

- Appreciate the physical positioning and attachment of the LVDT within a system to be affected by linear displacement.
- Have investigated the variable differential properties of the LVDT.
- Have produced a calibration graph for the LVDT and made judgements on such characteristics as: linearity, repeatability, accuracy, sensitivity.

Part (a): the variable properties of the LVDT

This experiment illustrates how the voltages of the secondary windings relate to the displacement of the soft iron core. It requires an oscilloscope, preferably dual beam. If you are not familiar with the use of an oscilloscope, refer to the oscilloscope manual or other source of instruction.

Ensure the whole linear assembly is moved to the extreme right and set to zero as indicated by the rotary scale.

Connect the circuit as shown in Figure 13.45. The circuit is represented schematically in Figure 13.46

Set the oscilloscope time base to 0.2 ms per division, and the gain to 2 volts per division. Adjust these settings during the experiment to get the best results.

Use the oscilloscope to measure the amplitude and frequency of the primary voltage and record your readings.

Measure the peak amplitude of the output voltage of secondary winding 1 over the complete range of displacement of the linear assembly. Tabulate your results as in Figure 13.47. At each step

observe the phase relationship between the primary and secondary signals.

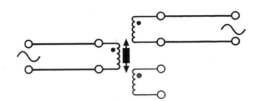

Figure 13.46 Schematic representation of the LVDT secondary winding 1 connected to an oscilloscope

Change the oscilloscope connection to secondary winding 2 and repeat the previous procedure.

On the same axes plot graphs of amplitude against displacement for the two secondary winding voltages.

Displacement (mm)	Secondary 1 (Peak V)	Secondary 2 (Peak V)
0		
1		
2		
3		
4		
5		
6		
7		
8		
9		

Figure 13.47 Results table

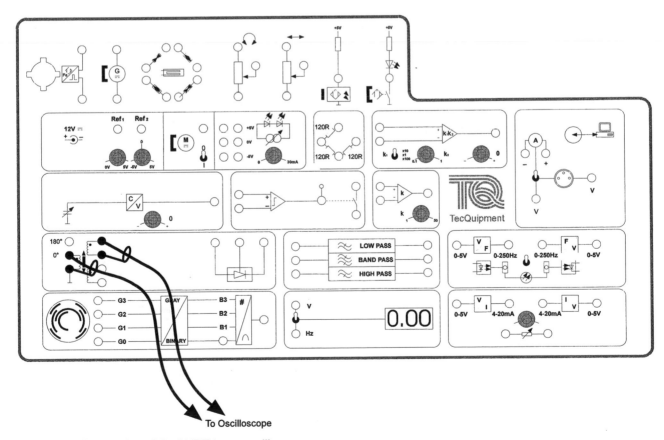

Figure 13.45 Connection of the LVDT to an oscilloscope

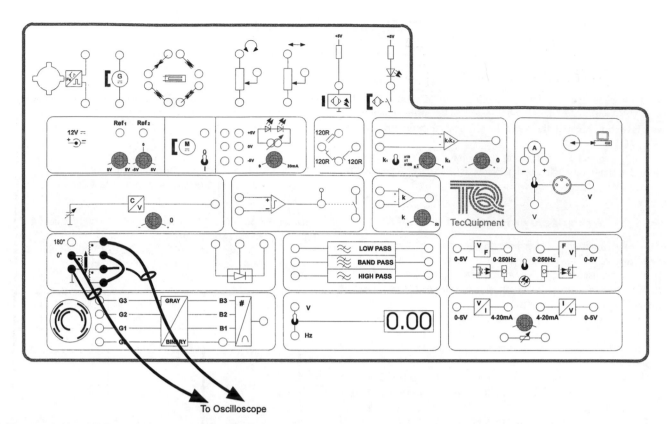

Figure 13.48 LVDT combined secondary winding connection diagram

Part (b): the differential properties of the LVDT

In this part of the experiment we look at how the voltage of a combined secondary winding varies with displacement.

Ensure the whole linear assembly is moved to the extreme right. Connect the circuit shown in Figure 13.48. This is represented schematically in Figure 13.49.

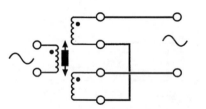

Figure 13.49 Schematic circuit of LVDT combined secondary winding

Displacement (mm)	Secondary Voltage (V)
0	
1	
2	
3	
4	
5	
6	
7	
8	
9	

Figure 13.50 Results table

Use the oscilloscope to measure the amplitude of the secondary voltage over the complete range of movement of the linear assembly. Tabulate your results as in Figure 13.50. At each step, observe the phase relationship between the primary winding waveform and the combined secondary winding waveform.

Plot a graph of amplitude against displacement from the results obtained.

Part (c): linear measurement with the LVDT

The circuit is shown schematically in Figure 13.51. Ensure the whole linear assembly is moved to the extreme right. Make the connections shown in Figure 13.52.

With the 0° reference connected to the phase sensitive detector, record meter readings against corresponding linear displacement settings over the range of movement of the linear assembly. Tabulate your results as in Figure 13.53. Repeat the procedure using the 180° reference.

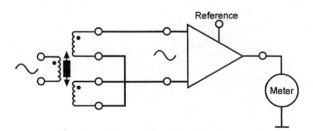

Figure 13.51 Schematic representation of LVDT linear measurement circuit

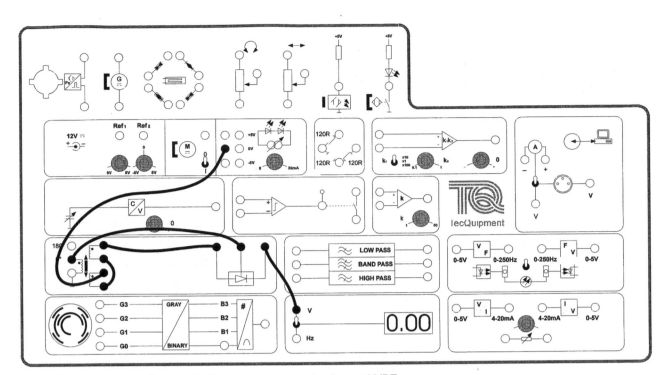

Figure 13.52 Connection diagram for linear measurement using an LVDT

Displacement (mm)	Output (V) 0°	Output (V) 180°
0		
1		
2		
3		
4		
5		
6		
7		
8		
9		

Figure 13.53 Results table

Plot a graph with displacement on the x-axis and LVDT voltage readings on the y-axis.

From the slope of the graph determine the sensitivity of the complete LVDT system. Compare this value with the LVDT specification.

Questions

Answer the following questions, making additional measurements if necessary. It may be helpful to refer to the theory given in Chapters 3, 9 and 10.

1. What was the shape of the graph of displacement against voltage for the complete LVDT system in Part (c)?
2. What were the values of the slope and intercepts of the graph?
3. What was the equation relating voltage to displacement for the complete LVDT system?
4. What is the significance of the intercepts? Should they be zero?
5. Was there any significant scatter of the results about the line of best fit drawn through the points?
6. What is your estimate of the error that would be introduced by the LVDT in normal use?
7. What is the sensitivity of the LVDT?
8. Which properties make the LVDT suitable for displacement measurement?
9. In which applications might LVDTs be used?

Experiment 4: the variable area capacitor

On completion of this experiment you will:

- Understand how a variable area capacitor can attach to a system to measure displacement.
- Have produced a calibration graph of output against displacement and made judgements on such characteristics as: linearity, repeatability, accuracy, sensitivity.
- Appreciate the sources of errors in capacitive circuits.

Part (a): the variable capacitor characteristics

The variable area capacitor on the TecQuipment SIS is connected to a capacitance to voltage converter. The front panel accesses the output from the capacitance to voltage converter.

Connect the equipment as shown in Figure 13.54.

Move the linear assembly to the right by rotating the manual control clockwise until it reaches the end stop. Carefully adjust the dial until the zero aligns with the edge of the moulding.

Use the 'set zero' control on the capacitance bridge output to zero the reading on the meter.

In steps of 1 mm (one complete rotation of the rotary scale), move the linear assembly to the left over its full range of travel. Record corresponding meter readings and tabulate your results as in Figure 13.55. Adjust the control in one direction only throughout the procedure.

Plot a graph of your results. Comment on the shape of the graph and measure its slope and intercept with the vertical axis. Hence give the equation which governs this measurement system.

With the linear assembly adjusted to be in midposition, determine the minimum amount of movement (resolution) that can be detected by the meter.

Displacement (mm)	Output (volts)
0	
1	
2	
3	
4	
5	
6	
7	
8	
9	

Figure 13.55 Results table

Part (b): variable capacitor with gain

Set the gain of the differential amplifier to 10. Connect both inputs of the amplifier together ('short' them). Connect the output of the amplifier to the DVM and adjust the set zero control until the meter reads zero. Do not change this setting throughout the rest of the experiment.

Connect the circuit as shown in Figure 13.56.

Use the 'set zero' control on the capacitance bridge to zero the reading on the meter once more. Repeat the previous procedure and tabulate your results. From these plot a graph.

Comment on the shape of the graph and measure its slope and intercept with the vertical axis. Hence give the equation which governs this measurement system.

With the linear assembly adjusted to be in midposition determine the minimum amount of movement (resolution) that can be detected by the meter.

Touch the body of the capacitor and note the effect this has on the meter reading. Connect the body of the capacitor to ground (0 volts) and again note the effect on the meter reading.

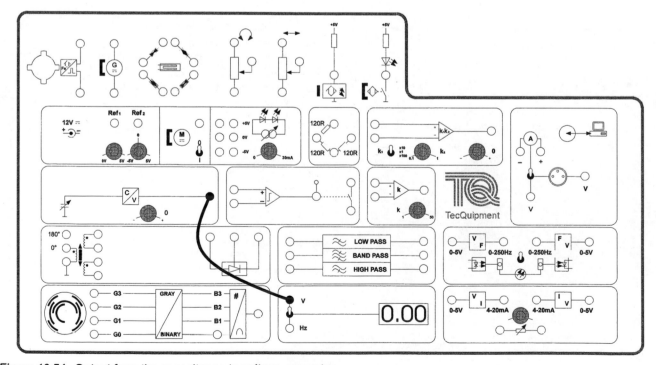

Figure 13.54 Output from the capacitance to voltage converter

Part (c): use of the comparator circuit

This experiment investigates the effect of using a comparator at the output of the capacitance system to achieve controlled switching.

The signals obtained in Parts (a) and (b) vary continuously with displacement to give a measure of displacement. This is in keeping with many types of sensor.

There are applications where some action is needed when a system reaches a predefined position, such as the limit stop on a piece of machinery. In others it is the movement of a body close to the probe which is sensed and a response needed. Touch switches use this principle. These use the change of capacitance when touched to initiate an action.

Using the results from Part (b), set Ref_1 to the same voltage obtained when the linear assembly was displaced 4 mm from the left.

Connect the circuit as shown in Figure 13.57.

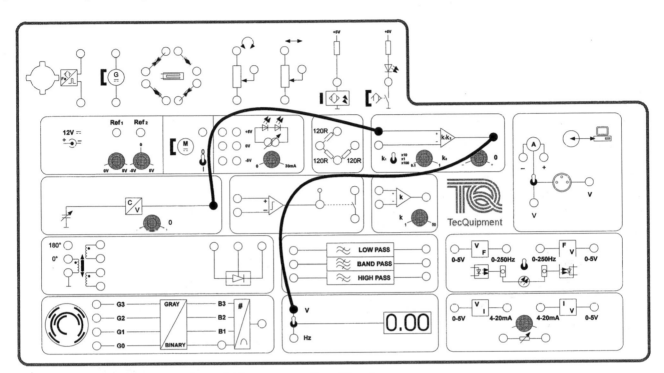

Figure 13.56 Connection diagram for variable area capacitor with gain

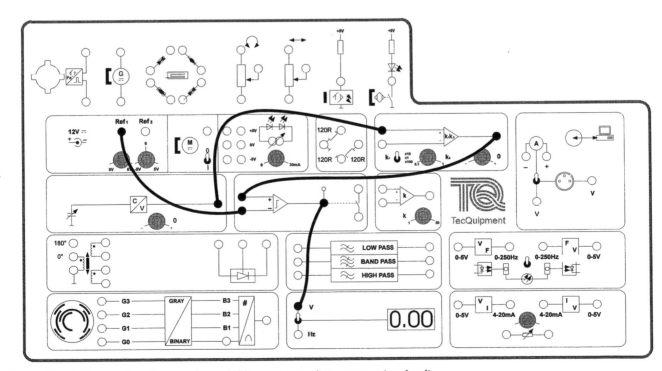

Figure 13.57 Connection diagram for variable area capacitor comparator circuit

Figure 13.58 shows this circuit schematically.

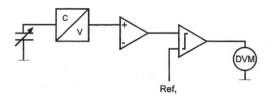

Figure 13.58 Variable area capacitor comparator circuit

Starting with the linear assembly to the right, move it to the left and observe the meter reading over the full range of movement.

Repeat the procedure with varying values of Ref_1 over its complete range.

Questions

Answer the following questions, making additional measurements if necessary. It may be helpful to refer to the theory given in Chapters 3, 9 and 10.

1. Describe how a variable area capacitor could be designed to measure rotary displacement.
2. With reference to the capacitor equation, describe other types of variable capacitor other than the variable area type. Give typical applications for the devices described and any practical precautions that need to be considered.
3. The maximum amount of relative movement between the inner and outer cylinders of the SIS variable area capacitor is 9 mm. If the internal diameter of the outer cylinder is 20.5 mm and the external diameter of the inner cylinder is 19.05 mm, calculate the change in capacitance that occurs when the linear assembly is moved over the complete range. State any assumptions that you have made.
4. Using the value calculated in question 3, and the results from Part (b) of this experiment, produce a calibration graph for the SIS variable area capacitor of displacement against capacitance.

Experiment 5: the reed switch

On completion of this experiment you will:

- Understand the functionality of the reed switch.
- Appreciate how and where a reed switch may be positioned within a system to measure proximity.
- Determine the resolution, repeatability and hysteresis of the reed switch system.
- Be aware of any limitations of the reed switch for measuring proximity.

This experiment investigates the operation of a reed switch in sensing the proximity of a magnet. Make the connection shown in Figure 13.59. Figure 13.60 shows this circuit schematically.

Figure 13.60 Schematic representation of the reed switch

Set the meter to read volts (V). Move the linear assembly to the right by rotating the manual control clockwise until it reaches the end stop. Carefully adjust the dial until the zero aligns with the edge of the moulding.

In steps of 1 mm move the linear assembly to the left over its full range of travel. Record corresponding meter readings and tabulate your results as in Figure 13.61. Adjust the control in one direction only throughout the procedure.

Where a contact closure occurs between readings return to that displacement range and, using smaller steps, determine the repeatability of the measurement, the resolution and the hysteresis.

Displacement (mm)	Output (volts)	Contacts open or closed
0		
1		
2		
3		
4		
5		
6		
7		
8		
9		

Figure 13.61 Results table

Questions

Answer the following questions, making additional measurements if necessary. It may be helpful to refer to the theory given in Chapters 3, 9 and 10.

1. Over the full range of movement of the linear assembly, how many times did the contact close? Explain why this was so.
2. How could the reed switch be mounted so that only one closure occurs for the same range of displacements?
3. In the SIS a permanent magnet initiates the reed switch action. What would be the effect if an electromagnet were to be used instead?
4. What would be the effect of either decreasing or increasing the strength of the magnet?

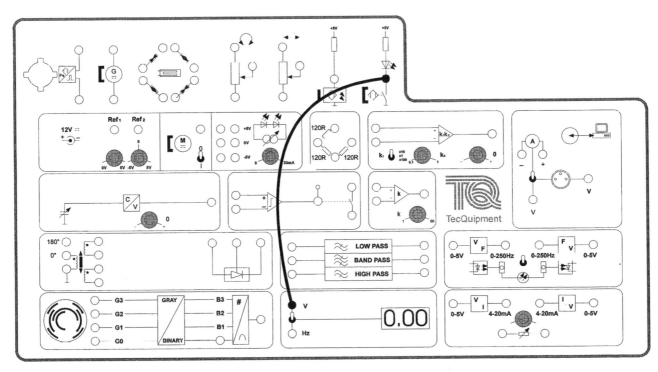

Figure 13.59 Reed switch connection diagram

Experiment 6: the reflective optical beam sensor

On completion of this experiment you will:

- Understand how a reflective optical beam sensor may be positioned within a system to measure proximity.
- Experience how to minimise the effects of background light on the sensitivity of the system.
- Appreciate the limitations of the reflective optical beam sensor for measuring displacement.

This experiment investigates the properties of the LED and phototransistor arrangement, and its use in sensing the proximity of the linear assembly.

Figure 13.62 shows the schematic arrangement between the LED and the phototransistor.

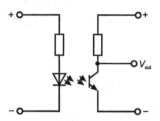

Figure 13.62 Schematic arrangement of the LED and phototransistor

Part (a): reflective optical sensor characteristics

Note that this experiment may be affected by the amount of background light surrounding the apparatus. If difficulties are found obtaining any change in output signal reduce the level of background light reaching the phototransistor.

Make the connection shown in Figure 13.63. Move the linear assembly to the right by turning the rotary scale clockwise until it reaches the end stop. Carefully adjust the dial until the zero aligns with the edge of the moulding.

Set the LED current level to maximum and the meter to read volts (V).

In steps of 1 mm move the linear assembly to the left over its full range of travel. Record corresponding meter readings and tabulate your results as in Figure 13.64. Adjust the control in one direction only throughout the procedure.

Repeat the procedure with the LED current setting at the '12 o'clock' position and then at the '9 o'clock' position. At each position observe the relative position of the end of the transverse beam and the output signal produced.

Displacement (mm)	Current at max. output (volts)	Output (volts) current at '12 o'clock'	Output (volts) current at '9 o'clock'
0			
1			
2			
3			
4			
5			
6			
7			
8			
9			

Figure 13.64 Results table

On the same axes plot graphs of meter reading against displacement for the three LED current levels.

Use your results to find the optimum setting for the LED current.

Examine the effect of changing the level of background light using a portable lamp or by moving the experiment to be closer to a window.

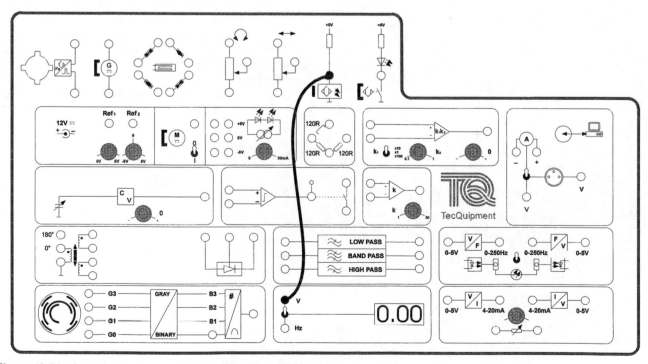

Figure 13.63 Connection diagram for the reflective optical sensor

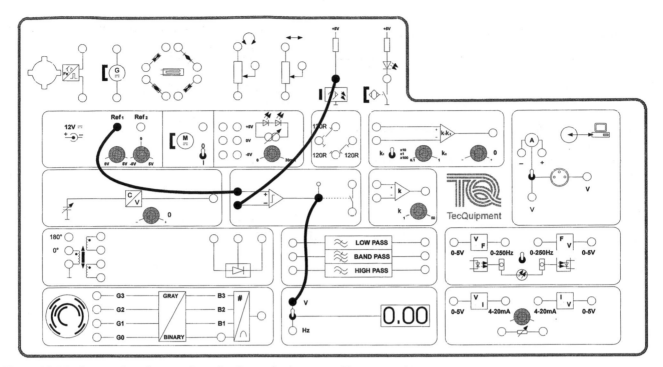

Figure 13.65   Connection diagram for reflective optical sensor with comparator

Part (b): using a comparator circuit

Modify the circuit to include the comparator at the output of the reflective optical beam sensor, as shown in Figure 13.65.

With the linear assembly moved to the right, adjust the setting of Ref_1 until the comparator indicator LED is off.

To investigate the effect on the reflective optical beam sensor of connecting the comparator to its output, repeat the previous procedure. Use the optimum LED current setting determined in Part (a). Tabulate your results as shown in Figure 13.66.

Displacement (mm)	Output (volts)
0	
1	
2	
3	
4	
5	
6	
7	
8	
9	

Figure 13.66   Results table

Determine the hysteresis and repeatability of the measurement system. Comment on your results and observations.

Questions

Answer the following questions, making additional measurements if necessary. It may be helpful to refer to the theory given in Chapters 3, 9 and 10.

1. What was the effect of varying the level of current flowing through the LED? What setting of the current control achieved optimum response? (It may be necessary to repeat the procedure at different current settings to obtain this value.)
2. What possible sources of error are there with this type of system? How could these errors be minimised?
3. Would you use an reflective optical system to measure the speed of rotation of a road wheel on a motor car? Give reasons for your answer.
4. What advantages are there in using a reflective optical beam sensor for proximity measurement compared with a reed switch.

Experiment 7: the optical tachometer

Complete this experiment before others which use the rotary section of the SIS. This is because it produces a calibrated standard.

On completion of this experiment you will have:

- Investigated the use of an optical sensor system with disc for speed measurement.
- Produced a calibration graph of rotary speed against input voltage.
- Identified sources of errors and loss of signal.

This experiment uses the least significant bit (LSB) of the SIS encoder disc. When a slot in the disc coincides with the LED and phototransistor arrangement, light passes through the disc from the LED to the phototransistor. This causes the transistor to conduct, lowering the output voltage, V_{out}. When the disc is rotated the level of light reaching the phototransistor alternates between high and low as the slots pass by. This gives a corresponding change in the output voltage.

Since the rate of change of the output signal state will vary with shaft speed, the frequency of the output from the phototransistor corresponds to the speed of rotation.

Part (a): the photo transmissive system

Figure 13.68 represents the LED and phototransistor arrangement schematically. Make the connections shown in Figure 13.67.

Rotate the current control found in the power supply section of the front panel fully clockwise.

Turn disc manually until the white indexing mark on the edge of the disc is pointing forwards. At this point LED's in the decoder section will be off.

Set the meter switch to read volts. Carefully rotate the disc clockwise (as viewed from the left-hand side) through 360° until the LED G0 lights up. Ignore all other LED's.

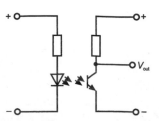

Figure 13.68 Schematic diagram of the LED and phototransistor arrangement

Record whether a slot or solid part of the disc is adjacent to the LED and phototransistor. Note whether LED G0 is lit up and also the indicated meter voltage. Tabulate your results as in Figure 13.69

Disc position	Slot or solid	LED G0 (on/off)	Output (volts)
0 (Index)			
1			
2			
3			
4			
5			
6			
7			
8 (Index)			

Figure 13.69 Results table

Continue to rotate the disc in the same direction and take readings at each change in the illumination state of the LED G0.

Repeat the procedure to complete the table and the indexing mark is once more facing forward. This represents one complete revolution of the disc or 2π radians (360°).

Determine the frequency of output signal for a given speed of rotation. Comment on your results.

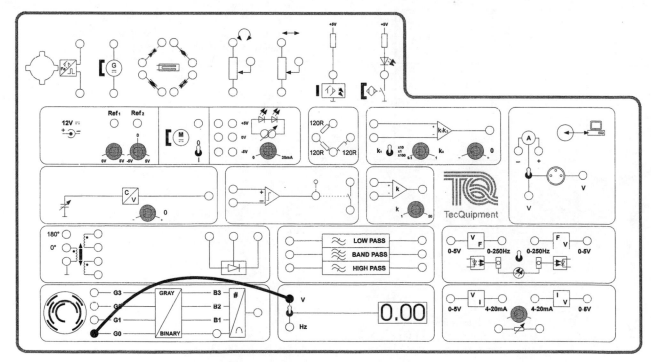

Figure 13.67 Connection diagram for the optical tachometer

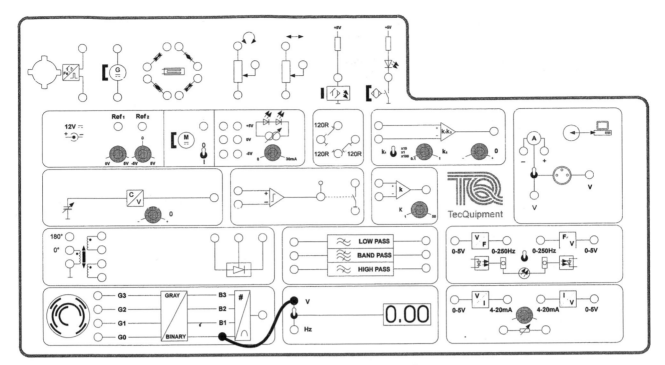

Figure 13.70 Connection diagram for optical tachometer output via Gray to binary converter

Part (b): the photo transmissive system and the Gray to binary converter

In this part of the experiment we use an alternative signal source to measure the output frequency from the optical tachometer. This is the LSB of a four-bit binary signal, B0, derived by the Gray to binary converter (this conversion is considered in more detail in Experiment 10).

Change the connection to the meter as shown in Figure 13.70.

Disc position	Slot or solid	LED B0 (lit or unlit)	Output (volts)
0 (Index)			
1			
2			
3			
4			
5			
6			
7			
8			
9			
10			
11			
12			
13			
14			
15			
16 (Index)			

Figure 13.71 Results table

Repeat the procedure given in Part (a), referring to LED B0 instead of G0. Tabulate your results as in Figure 13.71.

Determine the frequency of the output signal for a given speed of rotation Compare these results with those obtained in Part (a).

Calculate the frequency of the optical tachometer output signal at B0, for the speeds indicated by the table in Figure 13.72. Tabulate your results and keep the table for future reference. It will be used for other experiments. Hence plot a calibration graph of output frequency against speed of rotation in the range of 0 to 209.44 radians per second (0 to 2000 revolutions per minute). Comment on the results obtained and the calibration graph produced.

Speed (rev.min^{-1})	Speed (rad.s^{-1})	Frequency (Hz)
0	0	
200	20.94	
400	41.88	
600	62.83	
800	83.78	
1000	104.72	
1200	125.66	
1400	146.61	
1600	167.55	
1800	188.50	
2000	209.44	

Figure 13.72 Calibration table

Part (c): the motor speed characteristic

In this part of the experiment we use the calibration graph produced in Part (b) to investigate the speed characteristics of the d.c. motor driving the shaft assembly.

Set the motor drive control switch to 0. Connect the circuit as shown in Figure 13.73. Set the motor drive control switch to 1. Starting with the Ref$_2$ control fully anticlockwise, rotate the control clockwise in steps as indicated by the table in Figure 13.74.

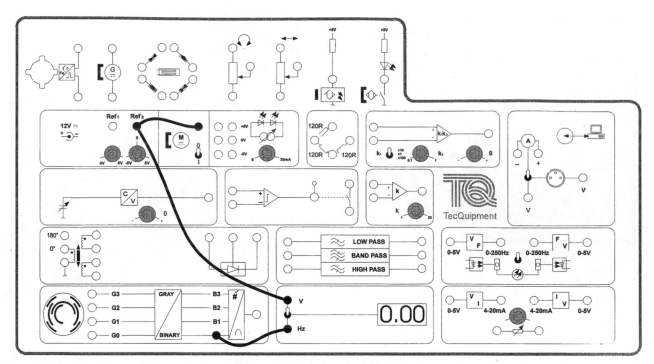

Figure 13.73 Connection diagram for investigating the speed characteristics

Record corresponding values of frequency and motor input voltage (use the meter toggle switch to obtain the voltage and frequency readings). Calculate the speed of rotation for each frequency to complete the table.

Input voltage (V)	Frequency (Hz)	Speed (rad.s^{-1})
−5.0		
−4.5		
−4.0		
−3.5		
−3.0		
−2.5		
−2.0		
−1.5		
−1.0		
−0.5		
0.0		
0.5		
1.0		
1.5		
2.0		
2.5		
3.0		
3.5		
4.0		
4.5		
5.0		

Figure 13.74 Results table

Plot a graph of motor input voltage against motor speed. Comment on the results obtained.

Questions

Answer the following questions, making additional measurements if necessary. It may be helpful to refer to the theory given in Chapters 3, 9 and 10.

1. How significant is the output frequency from the optical tachometer?
2. Can the optical tachometer measure the position of the disc?
3. If the slots in the disc are shortened what effect would this have on the results?
4. Can the SIS optical tachometer be used to determine the direction of rotation of the disc?
5. Will the accuracy of measurement be affected by the level of light emitted from the LED?
6. Would you expect the accuracy of the optical tachometer to drift over a period of time? If so, why?
7. Was it valid to calculate the values of speed and frequency to produce the calibration graph in Part (b)?
8. Describe an alternative optical technique that could be used for measuring speed of rotation. Are there any advantages or disadvantages with this technique compared with the system we have looked at?

Experiment 8: the d.c. tachometric generator

On completion of this experiment you will:

- Understand how a d.c. tachometric generator may be used to measure rotary speed.
- Have calibrated a d.c. tachometric generator and assessed the quality of measurement.
- Identify any limitations in the measurement system.

In this experiment we use the data produced in Experiment 7, Figure 13.72 to produce a calibration curve for the d.c. tachometric generator.

Switch the motor drive control switch to 0. Connect the circuit as shown in Figure 13.75.

Switch the motor drive control switch to 1. Starting with the Ref$_2$ control fully anticlockwise, rotate the control clockwise until the frequency indicated by the meter is the same as the first value in Figure 13.72 of Experiment 7. Set the indicated frequency to be as close as possible to minimise errors.

Record the corresponding value of tachometric generator output voltage. It will be necessary to switch between the meter frequency and voltage ranges to obtain both readings.

Repeat the procedure over the full range of Ref$_2$, tabulating your results as in Figure 13.76.

Remember to allow for the change in direction of rotation when passing through zero volts at the input to the motor drive circuit. Use negative speeds for negative applied voltages and positive speeds for positive voltages.

Plot a graph of tachometric generator output voltage against speed of rotation.

Comment on your results, especially the accuracy of the procedure.

Speed (rad.s^{-1})	Frequency (Hz)	Output (V)
(Insert values obtained in Experiment 7)		

Figure 13.76 Results table

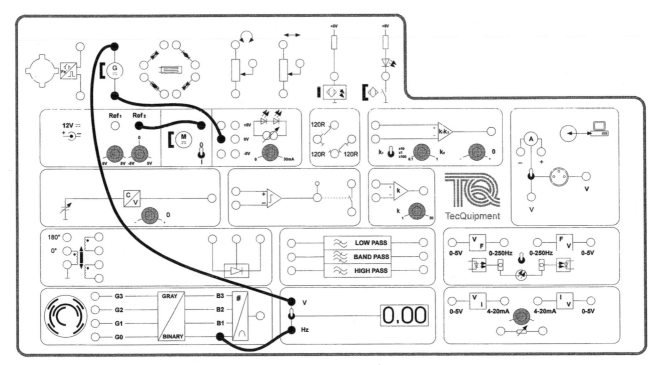

Figure 13.75 Connection diagram for the d.c. tachometric generator

Questions

Answer the following questions, making additional measurements if necessary. It may be helpful to refer to the theory given in Chapters 3, 9 and 10.

1. Can the d.c. tachometric generator be used to measure the position of the disc?
2. Can the d.c. tachometric generator be used to determine the direction of rotation?
3. Would you expect the accuracy of the d.c. tachometric generator to drift over a period of time? If so, why?
4. Describe an alternative type of tachometric generator that could be used for measuring speed of rotation. Are there any advantages or disadvantages with this technique compared with the d.c. tachometric generator used here?

Experiment 9: the variable reluctance probe

On completion of this experiment you will,

- Understand how a variable reluctance probe may be used to measure speed.
- Appreciate the need for amplification.
- Identify the effect of noise on a signal and its removal.

This experiment requires an oscilloscope.

Part (a): variable reluctance probe operation

Set the motor control switch to 0 and Ref_1 to minimum (fully anticlockwise).

This circuit is shown schematically in Figure 13.77. Make the connections shown in Figure 13.78.

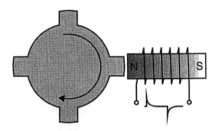

Figure 13.77   Variable reluctance probe

Set the motor drive control switch to 1. Adjust Ref_1 to the point where the motor is just starting to rotate smoothly.

Use the oscilloscope to monitor the waveform of the output signal from the variable reluctance probe. Adjust the oscilloscope settings to suit the waveforms produced at each step. Suitable initial values are 10 ms/div. and 20 mV/div.

Note the effect on the output signal of increasing Ref_1 to maximum (5 V).

Sketch, the waveform when Ref_1 is 5 V. Comment on your observations.

Part (b): calibration and speed measurement

From Part (a) you should have seen that the output signal from the variable reluctance probe is dependent upon the speed of rotation, in terms of both amplitude and frequency.

In this part of the experiment we use the relationship between frequency and speed to produce a calibration curve.

Set Ref_1 to minimum and the motor drive control switch to 0. Make the connections shown in Figure 13.79. Set the motor control switch to 1.

Increase Ref_1 in steps, as indicated by the table in Figure 13.80. Record the corresponding value of frequency (it will be necessary to switch between the voltage and frequency ranges of the meter). If the readings are erratic make the best estimation of the mid range value.

For each frequency calculate the corresponding value of speed, based upon there being 4 slots in the path of the probe. Compare your values of calculated speed against supply voltage with those obtained in Experiment 7 (Figure 13.74).

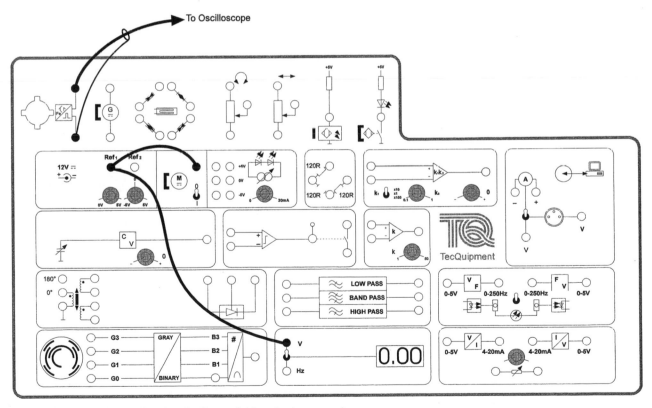

Figure 13.78   Connection diagram for the variable reluctance probe

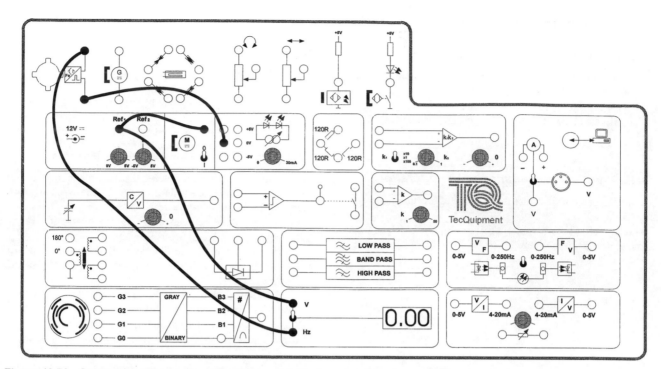

Figure 13.79 Connection diagram for calibration and speed measurement

Ref₁ (V)	Frequency (Hz)	Speed (rad.s⁻¹)
0		
0.5		
1.0		
1.5		
2.0		
2.5		
3.0		
3.5		
4.0		
4.5		
5.0		

Figure 13.80 Variable reluctance probe calibration table

Ref₁ (V)	Frequency at gain = 10 (Hz)	Frequency at gain = 100 (Hz)
0		
0.5		
1.0		
1.5		
2.0		
2.5		
3.0		
3.5		
4.0		
4.5		
5.0		

Figure 13.81 Results table

Part (c): calibration and speed measurement with amplified output

Part (b) showed that at higher speeds the probe was able to accurately measure frequency, compared with the results obtained in Experiment 7.

At low speeds however this was not the case. The low amplitudes of the signals could not be recognised by the meter circuit and, although the disc was rotating, no frequencies were displayed.

The object of this part of the experiment is to amplify the output from the variable reluctance probe. This will extend the range of measurement to suit the display instrument.

Set the motor drive control switch to 0. Zero the output from the differential amplifier with its inputs shorted together and set the gain to 10.

Make the connections shown in Figure 13.82. Set Ref₁ to minimum (fully anticlockwise) and the motor drive control switch to 1.

Repeat the procedure in Part (b) and tabulate your results as in Figure 13.81. Repeat the procedure with the differential amplifier gain set to 100 (there is no need to calculate the speed at each reading). Compare your results with those obtained in Part (b).

Part (d): calibration and speed measurement incorporating a low pass filter

Part (c) showed that with the gain set to 10 there was a small improvement in the lower speed measurement of frequency. With the gain set to 100 the readings were erratic. The main reason for this is noise present in the system.

Again connect the circuit as shown in Figure 13.82. Use an oscilloscope to compare the output signal from the variable reluctance probe and then the frequency input to the meter.

The object of this experiment is to use a low pass filter to remove high frequency noise and so improve the quality and hence accuracy of the measured signals.

Set the motor drive control switch to 0. Make the connections shown in Figure 13.83.

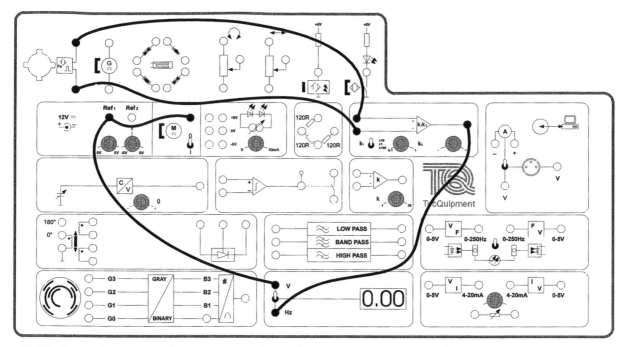

Figure 13.82 Connection diagram for calibration and speed measurement with an amplified output

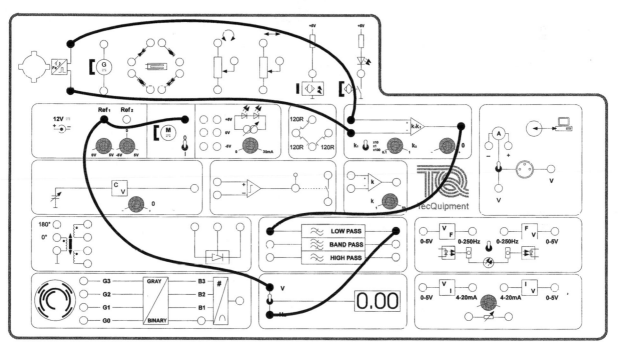

Figure 13.83 Connection diagram for calibration and speed measurement using a low pass filter

Ref₁ (V)	Frequency (Hz)
0	
0.5	
1.0	
1.5	
2.0	
2.5	
3.0	
3.5	
4.0	
4.5	
5.0	

Figure 13.84 Results table

Set the motor drive control switch to 1 and the differential amplifier gain to 100. Repeat the procedure in Part (b) and tabulate your results as in Figure 13.84.

Compare your results with those obtained in previous parts of this experiment, and also Experiment 7.

Questions

Answer the following questions, making additional measurements if necessary. It may be helpful to refer to the theory given in Chapters 3, 9 and 10.

1. How significant is the output frequency from the variable reluctance probe?

2. Can the variable reluctance probe be used to measure the position of the disc?

3. If the teeth in the disc are narrowed, what effect would this have on the results?

4. Can the variable reluctance probe determine the direction of rotation?

5. Will the accuracy of the measurement be affected by the distance between the disc and the probe?

6. Would you expect the accuracy of the variable reluctance probe to drift over a period of time? If so, why?

7. How could the variable reluctance probe measure linear speed?

Experiment 10: the four-bit optical encoder

On completion of this experiment you will:

- Understand how optical encoders measure displacement and derive speed.
- Have produced a table of rotary position with respect to a Gray scale and binary scale.
- Have made judgements on such characteristics as linearity, repeatability, accuracy, sensitivity.
- Recognise the sources of errors in encoders.

Part (a): the Gray scale

This part of the experiment looks into rotary position measurement using the Gray scale encoder. No extra circuit connections are needed (they exist in the internal circuitry of the hardware module). The four bits at the input to the decoder section are taken direct from the phototransistors monitoring the slots in the encoder disc.

Rotate the current control in the power supply section fully clockwise. This sets the current flowing through the LED to maximum.

Manually rotate the encoder disc until the white indexing mark on the edge of the disc points forwards. At this point all the LED's in the decoder section will be off. This is the starting position, Position 0. Rotate the disc clockwise, as viewed from the left-hand side of the hardware module. Record when a Gray scale LED (G0 to G3) lights up, in a table as shown in Figure 13.85. Record the on state with a large cross or shade in the cell to show the LED is on. This will show the coding pattern of the whole encoder when completed.

When the table is complete comment on the Gray coded scale.

Part (b): the binary scale encoder

This part of the experiment produces the binary scale for a four-bit encoder.

All four bits of the Gray scale are processed by a Gray to binary converter to produce a four-bit binary number, indicated by LED's B0 to B3. This is the same binary number that would be obtained if the encoder disc was binary coded.

Rotate the disc to position 0. As in part (a), rotate the disc manually, this time recording the state of the binary LED's (B0 to B3). Tabulate your results as in Figure 13.86. Comment on the binary coded scale when completed.

Part (c): digital to analogue conversion

The four bits of the binary number are processed by a digital to analogue converter. This provides an analogue signal which corresponds to the value of the digital number and so also indicates the disc position.

This part of the experiment investigates the analogue signal obtained using a four-bit encoder and decoder, and digital to analogue converter.

Make the connection shown in Figure 13.87. Set the meter to read volts (V).

Manually rotate the disc to the starting position, Position 0. Rotate the disc clockwise, as viewed from the left-hand side of the hardware module. Each time an LED illuminates record the meter reading. Tabulate your results as in Figure 13.88.

Plot a graph of position against meter reading. Comment on your results.

Position	0	1	2	3	4	5	6	7	8	9	10	11	12	13	14	15
G0																
G1																
G2																
G3																

Figure 13.85 Gray scale encoder output

Position	0	1	2	3	4	5	6	7	8	9	10	11	12	13	14	15
B0																
B1																
B2																
B3																

Figure 13.86 Gray to binary encoder output

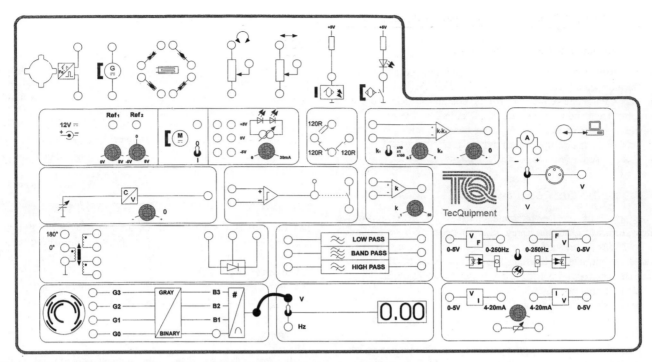

Figure 13.87 Digital to analogue conversion connection diagram

Disc position	Meter reading (V)
0	
1	
2	
3	
4	
5	
6	
7	
8	
9	
10	
11	
12	
13	
14	
15	
16	

Figure 13.88 Results table

Questions

Answer the following questions, making additional measurements if necessary. It may be helpful to refer to the theory given in Chapters 3, 9 and 10.

1. What is the effect of increasing the number of bits in the digital encoder? Are there any limitations of how many bits are possible?
2. What is the resolution of an 8 bit encoder?
3. How can the encoder be used to measure speed of rotation?
4. What alternative sensors are available for measuring shaft position if the rotation is to be less than 360°? What are the advantages and disadvantages of these alternatives?

Experiment 11: data transmission

On completion of this experiment you will:

- Appreciate the sources of errors and loss of signals when sending data between a source and a receiver.
- Experience the reduction in errors using voltage to frequency and frequency to voltage and current loop transmission techniques.
- Be introduced to data transmission using fibre optic techniques.

Part (a): conventional data transmission

This part of the experiment provides a reference against which the later techniques will be compared.

We will simulate a control signal sent from a source to a receiver, in this case a motor. Assume the connection between the source and the receiver is a single long cable, shown schematically in Figure 13.89.

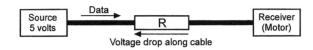

Figure 13.89 Schematic representation of a data transmission signal

The resistance of the cable is shown as a resistance of value R. In this experiment the resistance used is a variable resistor.

Set the motor drive control switch to 0. Set Ref_1 to minimum, the meter to read voltage (V) and the variable resistor in the V to I and I to V section to minimum (both fully anticlockwise).

Make the connections shown in Figure 13.90. The free end of the meter lead will be used to make measurements at more than one point in the circuit.

Set the motor drive control switch to 1. Increase Ref_1 to maximum. Measure the output voltage and then the input voltage to the motor drive circuit.

Repeat the procedure with the variable resistor at mid position, and then at maximum. Tabulate your readings as in Figure 13.91. Comment on your results.

Meter reading (volts)	Minimum resistance (volts)	Mid Position resistance (volts)	Maximum resistance (volts)
At Ref_1			
At motor			

Figure 13.91 Results table

Part (b): current loop transmission

In this part of the experiment we include a voltage to current (V to I) converter before the variable resistance and a current to voltage (I to V) converter after the variable resistance. This is to investigate their effect on signal weakening (attenuation).

Set the motor drive control switch to 0 and Ref_1 to minimum. Make the connections shown in Figure 13.92.

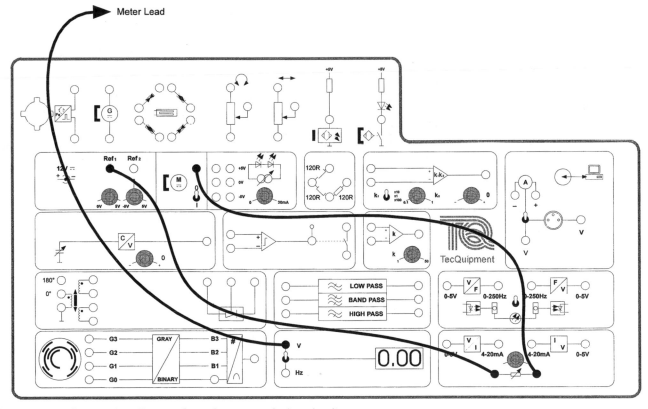

Figure 13.90 Connection diagram for a data transmission circuit

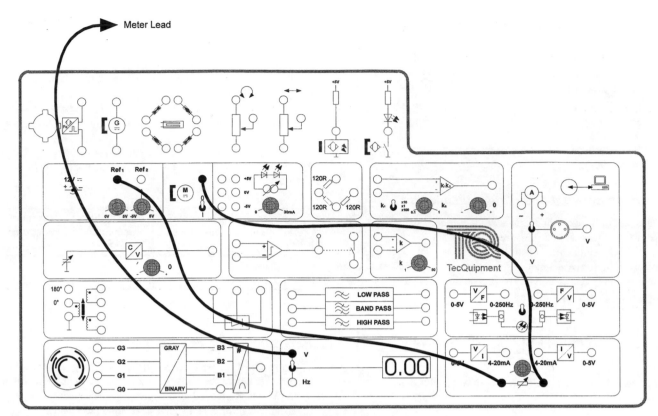

Figure 13.92 Connection diagram for current loop transmission

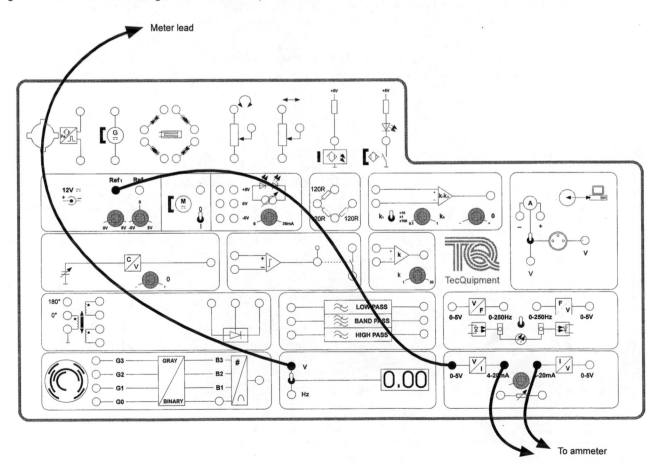

Figure 13.93 Calibration connection diagram

Repeat the procedure of Part (a). Tabulate and comment on your results.

To find the linearity of the process, we will now produce a calibration graph for both converters. This part of the experiment needs an ammeter rated up to 25 mA, or suitable computer software.

Set Ref_1 to minimum. Make the circuit connection shown in Figure 13.93.

V to I input voltage (V)	V to I output current (mA)	I to V output voltage (V)
0.0		
0.5		
1.0		
1.5		
2.0		
2.5		
3.0		
3.5		
4.0		
4.5		
5.0		

Figure 13.94 Calibration table

In steps, as indicated by the table in Figure 13.94, increase the value of Ref_1. Record corresponding values of the input voltage and output current from the voltage to current converter. Also record the output voltage of the current to voltage converter. The output current of the current to voltage converter is the same as the input current to the current to voltage converter.

Plot a graph of voltage to current converter input voltage against output current. Plot a separate graph of current to voltage converter input current against output voltage. Plot a third graph

of voltage to current converter input voltage against current to voltage converter output voltage. Comment on your results.

Part (c): digital transmission

In this part of the experiment we include a voltage to frequency (V to F) converter before the variable resistance and a frequency to voltage (F to V) converter after the variable resistance. Again this is to investigate their effect on signal attenuation.

Set the motor drive control switch to 0, Ref_1 to minimum, the meter to read voltage and the variable resistance to minimum. Make the connections shown in Figure 13.95.

Set the motor drive control switch to 1. Increase Ref_1 to maximum. Measure the output voltage and then the input voltage to the motor drive circuit. Repeat the procedure with the variable resistor at mid position, and then at maximum. Tabulate your results as in Figure 13.96.

Meter reading (V)	Minimum resistance (V)	Mid position resistance (V)	Maximum resistance (V)
At Ref_1			
At Motor			

Figure 13.96 Results table

Comment on your results. To find the linearity of the process, we will now produce a calibration graph for both converters. Set Ref_1 to minimum and the fibre optic selector switch up (away from you).

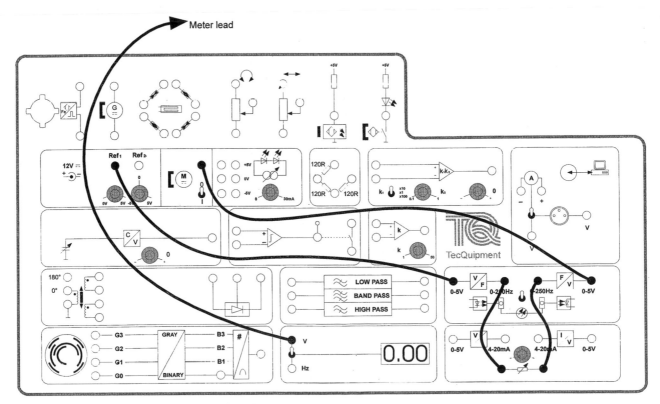

Figure 13.95 Digital transmission connection diagram

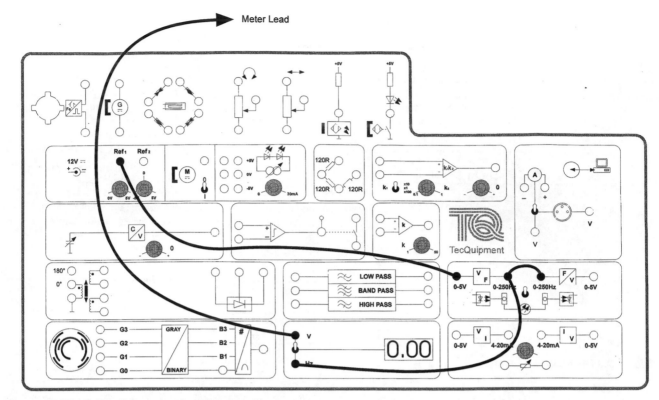

Figure 13.97 Calibration connection diagram

Make the connections shown in Figure 13.97. In steps, as indicated by the table in Figure 13.98, increase the value of Ref$_1$. Record corresponding values of the input voltage and output frequency from the voltage to frequency converter. Also record the output voltage of the frequency to voltage converter.

V to F input voltage (volts)	V to F output frequency (Hz)	F to V output voltage (volts)
0.0		
0.5		
1.0		
1.5		
2.0		
2.5		
3.0		
3.5		
4.0		
4.5		
5.0		

Figure 13.98 Calibration table

The output frequency of the frequency to voltage converter is the same as the input frequency to the frequency to voltage converter.

Plot a graph of voltage to frequency converter input voltage against output frequency. Plot a separate graph of frequency to voltage converter input frequency against output voltage. Plot a third graph of voltage to frequency converter input voltage against frequency to voltage converter output voltage. Comment on your results.

Part (d): fibre optic transmission

This part of the experiment investigates the effect of fibre optic transmission on signal attenuation.

Make the connections shown in Figure 13.99. Make the link between the voltage to frequency and frequency to voltage converters using an optical fibre. Insert the ends of the fibre optic cable, terminated at a plug, into the corresponding sockets in the V to F/F to V section of the front panel.

Set the fibre optic selector switch in the V to F/F to V section of the front panel down (towards you). This switches in the fibre optic link, including an LED at the sending end and a phototransistor at the receiving end. The access sockets are still available to measure the frequency at the output from the voltage to frequency and the input to the frequency to voltage converters.

Repeat the procedure in Part (c). Tabulate your results as in Figure 13.100. Compare your results with those obtained in Part (c). Comment on your results.

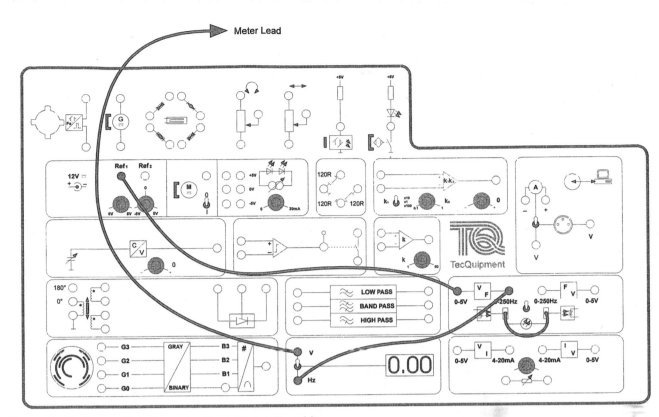

Figure 13.99   Connection diagram using a fibre optic cable

V to F input voltage (volts)	V to F output frequency (Hz)	F to V output voltage (volts)
0.0		
0.5		
1.0		
1.5		
2.0		
2.5		
3.0		
3.5		
4.0		
4.5		
5.0		

Figure 13.100   Calibration table (with fibre optic cable)

Questions

Answer the following questions, making additional measurements if necessary. It may be helpful to refer to the theory given in Chapters 3, 9 and 10.

1. In Part (d), why was the effect of using the variable resistor not investigated?
2. In Part (d), with Ref_1 at 5 volts, what is the effect of removing the fibre optic cable (just pull one end out if its socket)? As you replace the fibre back into its socket, what is the effect as the distance reduces and alignment improves?
3. Using an oscilloscope, what would you expect to see at the output of the voltage to frequency converter, and the frequency to voltage converter in Part (c) if Ref_1 was varied from 0 to 5 volts?

Experiment 12: introduction to control

This experiment investigates the principles of open- and closed-loop control of a rotating body using, a d.c. tachometric generator to provide feedback. On completion of this experiment you will have:

- Investigated open-loop and closed-loop circuits for control of speed.
- Experienced the concepts of set point, error and control signal.

Part (a): open-loop control

In this part of the experiment a control signal is set manually to produce the speed required.

Set the meter to read voltage (V). This display will be used to monitor speed. Set the motor drive control switch to 0 and adjust Ref_1 to produce an indicated value of 2.5 volts.

Make the connections shown in Figure 13.101. Set the motor control switch to 1. Use the meter to measure the output voltage from the tachometric generator and compare this with the value of Ref_1. Calculate the difference or error.

Momentarily apply a load to the motor by pressing the push-button adjacent to the tachometric generator legend. Note the

effect on the speed. Determine the value of the error, and comment on your results.

Part (b): closed-loop control

In this experiment the output variable is the speed of the d.c. motor. The output from the d.c. tachometric generator, the feedback signal, is compared with the reference voltage Ref_1 by the summing amplifier. This produces the error signal. The error signal is amplified by the controller to become the control signal. The control signal drives the motor and so determines the speed of rotation.

Set the motor drive switch to 0. Adjust Ref_1 to 2.5 volts and the differencing amplifier gain to minimum (1) by rotating the control fully anticlockwise.

Make the connections shown in Figure 13.102. Set the motor drive switch to 1. With the gain at (1) the error signal is supplied directly to the motor drive circuit.

Note the value of the tachometric generator signal indicated by the meter. How does this compare with the value from Part (a)? Note what happens to the error signal when the motor is loaded. Explain why this is so and comment on your observations.

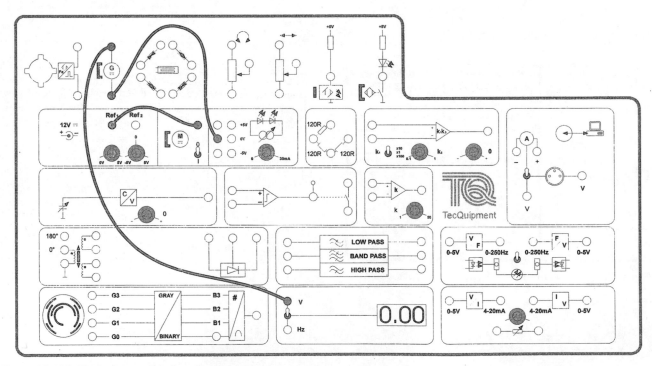

Figure 13.101 Open-loop control connection diagram

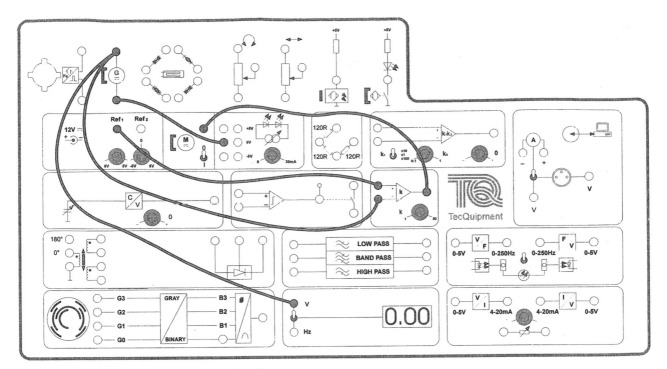

Figure 13.102 Closed-loop control connection diagram

Questions

1. In a closed-loop control system, can the error ever be zero?
2. What problems would you anticipate if you used the output signal from the digital to analogue converter as the feedback signal to control rotary position?
3. A reservoir of hot water in a process is to be maintained at a constant temperature. Any water drained from the tank is automatically replaced from a supply at a lower temperature. Describe a controller system which would maintain the temperature of water at the constant temperature with minimum error. State the devices you would use and why you have selected them. Include any considerations in the system design and component selection.

Summary

This chapter contains a structured series of experiments which illustrate the practical use and characteristics of a number of sensors. This is to support the theory given previously.

The experiments are ideally performed using the TecQuipment Sensors and Instrumentation System (SIS) hardware, but may be performed on similar equipment if available.

Further experimentation is possible, for example looking at proportional, integral and differential (PID) control. Also, the scope of the experiments can be expanded by interfacing to a PC and incorporating data acquisition software.

Because of variations in sensor specifications, characteristics and manufacturing tolerances, it has not been possible to provide experimental results here. However, if the experiments have been performed using the TecQuipment SIS hardware, typical results to all these experiments and answers to the questions in this chapter are given in the lecturer's guide which accompanies the product. Also supplied with the TecQuipment SIS are a student guide, an interactive CD-ROM, and automatic data acquisition software, making it a complete course in sensors and instrumentation.

Index

absolute
    measurement 6
    pressure 35
    zero 45
acceleration 24
accelerometer 24–5, 98
actuator 106
accuracy 7
amplifier 85
    analogue-to-digital converter 92
    charge 25
    current-to-voltage converter 91, 118–19
    differential 88, 117
    differentiating 90
    digital-to-analogue converter 92–3
    frequency-to-voltage converter 91–92, 118
    gain 85
    instrumentation 88
    integrating 88–9
    inverting 86
    non-inverting 87
    operational 85
    summing 87–8, 118
    voltage comparator 86, 133–4, 117, 137
    voltage follower 91, 118–19
    voltage-to-current converter 91, 118–19
    voltage-to-frequency converter 91, 118
analogue
    display 67–70, 105
    signal 67
angular displacement 12, 19-21, 24
anti-lock brakes 100–102

balance and scales 33
barometer 42–3
    aneroid 42–3
    liquid 42
bellows 39–40
bimetallic strip 46–7
    thermometer 47
    thermostat 47
Bourdon tube 38–9
bubbler level gauge 30–31

calibration 7
capacitance 17
    probe 29–30
capacitance-to-voltage converter 117
capacitive displacement transducer 17–19

capacitive pressure sensor 40–41
capacitor 17
chart recorder 73–4
circuits
    active 85
    passive 77
closed-loop control 5–6, 154
coefficient of resistance 48
conditioned signal 6
conductance probe 30
constriction effect 61
counterweight 28

data
    acquisition 75–6, 119, 155
    logger 67
'dead-band' or 'dead-zone' 7–8
dial test indicator 12
differential pressure 35
digital
    display 70–73
    signals 67
dipstick 28
disappearing filament 52
displacement 12
    angular 19–21
    capacitive 17–19
    linear 12–19
    rotary 19–22
display
    analogue 67–70
    digital 70–73
    light emitting diode 20–23, 71–72
    liquid crystal 72
    mechanical digital 72
    seven-segment 71
    sixteen-segment 71
Doppler effect 64

e.m.f 6
earth tremors 25
elastic pressure sensors 38–42
electrical
    noise 7
    resistance 48, 68–9
electromagnetic flow meter 63–4
emissivity 52
encode
    absolute 20–21

four-bit 116–17, 147–8
  incremental 19–20
  optical shaft 19–21

Faraday's law 13
fibre optics 152–3
filter 118, 144–5
float gauge 28
  counterweight 28
  electrical 28–9
flow meter
  electromagnetic 63–4
  helical screw 56
  hot-wire anemometer 60
  orifice plate 62
  paddle wheel 58
  pilot-static 59–60
  rotating lobe 56
  turbine 57–8
  ultrasonic 64
  variable area 60
  Venturi nozzle 63
  Venturi tube 61–2
flux 6
force 31–4
four-bit
  decoder 118, 147–8
  encoded 116–17, 147–8

galavanometric recorder 73
gas-filled tube 72–3
gauge
  Bourdon 38–9, 46
  bubbler level 30–31
  factor 15–16
  float operated 28–9
  pressure 35
  strain 15, 41–2
  strain, bonded resistance 15, 17
gravimetric tank 58–9
gray
  code 21
  scale 147

Hall effect 23
  sensor 23
Hooke's law 34
hot-wire anemometer 60
hysteresis 8

impedance 83
  matching 83–4
inductance 81
infra-red pyrometer 53

kelvin 45
Kirchhoff's law 78

light emitting diode 20
linear displacement 12–19
linearity 8
liquid crystal 72
load cell 31–3
LVDT 13–15, 115, 129–31

manometer 35–8, 42
  inclined tube 37–8
  U-tube 35–6
mass flow 58–9
measurand 6
measurement
  absolute 6
  non-disruptive flow 63–4
  system 3–4
measuring cylinder 27
meter
  moving coil 68–9
  moving iron 69
  turbine 57–8
  ultrasonic flow 64
microstrain 16
microswitch 22
moving coil recorder 73

nozzle 62–3
  Venturi 63

open-loop control 4–5, 154
operational amplifier 85
  gain 85
optical sensor 23, 155–16, 136–7
optical shaft encoder 19–21
orifice plate 62
oscilloscope 69–70

permittivity 17
phase sensitive detector 15, 117–18
  rectifier 108
phototransistor 23, 136, 138
piezoelectric
  accelerator 25
  effect 25
  pressure sensor 41
piezoresistive pressure sensor 41–2
pilot tube 59–60
pneumatic loop 98–9
potential divider 78, 120–21
potentiometer 12
  helical 19
  linear 12–13, 115, 125–6
  rotary 19, 115, 127
power matching 82–3
pressure 35
  absolute 35
  gauge 35
probe 7
  capacitance 29–30
  conductance 30, 63
  variable 2, 117, 143–6
proximity sensor 2–4, 117
pyrometer 45, 52–3
  disappearing filament 52
  infra-red 53

radian 11
reed switch 24, 115, 135
  sensor 24
relative permittivity 17
resistance 48

rotameter 60
rotating lobe meter 56

Seeback effect 50
seismic mass accelerometer 24–5
sensor characteristics 7–9
shock 24
SI units 7
sight glass 28
signal conditioning 117
spring balance 34
strain gauge 31–2, 41–2, 115, 120–4
    bonded resistance 15–17
    semiconductor 41
stress 35

tachometer 21
    optical 117, 138–9
tachometric generator 21, 116, 141
temperature coefficient of resistance 48
terminology 6–7
thermal array recorder 75
thermistor 49–50
thermocouple 51
thermometer
    bimetallic 47
    electrical resistance 48–9
    liquid-in-glass 45–6
    liquid-in-metal 46
thermopile 51

thermostat 47
tolerance 9
transducer 3
    capacitive displacement 18–19
    pressure 41
transformer 13
    linear variable differential 13–15, 115, 129–131

ultrasonic flow meter 64
    level indicator 30
ultraviolet light recorder 74–5

variable area capacitor 17–19, 115, 132–4
variable reluctance proximity sensor 22, 117, 143–6
velocity
    angular 1
    linear 11
Venturi nozzle 63
    tube 61
vibration 24–5, 97–100
virtual voltmeter 75
volt 6
volumetric flow rate 55–8

weight 31–4
Wheatstone bridge 16, 32, 80–81
window comparator 97

XY plotter 74